The Geology of Letchworth State Park:
Grand Canyon of the East

By William A. Szary

Copyright 2019. Earth2Energy. All Rights Reserved.

Book Cover: Front" A series of waterfalls are but a few of the attractions in the park. The wall rock consists of massive sandstone at the base overlain by bedded sandstone belonging to the Genesee Group. Back cover is the wall rock between the lower and middle falls belonging to the Lockport Dolomite unit of the Lower Silurian Period.

Library of Congress Catalog in Publications Data:

Szary, William A. The Geology of Letchworth State Park:
Grand Canyon of the East

Includes references

ISBN 13: 9781709938818

Earth2Energy Educational Publishing
Port Richey FL 34668

Earth2Energy is a Registered Trademark

Preface

The concept for this booklet was based on the scant geologic information available to the public concerning Letchworth State Park and the regional geology surrounding the park. A very limited number of publications were available for review about the technical aspects of the park and region. Most publications posted on the internet were of insufficient quality for the earth science student or geological student but were extremely adequate for the average tourist seeking information on the attractions in the park and region.

This booklet was an effort to compile and assemble technical geological information about the park and surrounding region based on what limited information could be acquired from the internet and based on a park visit that occurred during the Fall of 1983 before the park was fully developed into a tourist attraction. Based on the lack of information, it was deemed necessary to publish a technical effort in an attempt to explain the geology for the region and for Letchworth State Park.

Table of Contents

Preface 3

Chapter 1. Introduction 7

Chapter 2. General Geology 12
Geomorphology
Physiographic History
History of the Genesee River
Upper Gorge – Portageville to Mount Morris
Mount Morris to Irondeqoit Bay
Avon to Rochester

Chapter 3. Stratigraphic Succession 20
Ordovician System
Queenston Shale
Ordovician-Silurian Boundary

Silurian System
Lower Silurian
Grimsby Sandstone
Middle Silurian
Clinton Group
Thorold sandstone
Maplewood Shale
Reynales Formation
Rochester Shale
Lockport Dolomite
Upper Silurian
Salina Formation
Bertie Formation
Cobleskill Formation
Silurian Depositional History
Silurian-Devonian Hiatus

Devonian System
Lower or Middle Devonian
Onondega Limestone
Middle and Upper Devonian
Hamilton Group and Tully Formation
Marcellus Formation
Skaneateles Formation
Ludlowville Formation
Moscow Formation

Upper Devonian
Genesee Group
Geneseo Black Shale
Genundewa Limestone
West River Shale
Sedimentary Structures in the Genesee Group

Naples Group
Middlesex Black Shale
Cashaqua Formation
Rhinestreet Black Shale
Hatch Formation

Formations above the Naples Group
Grimes Sandstone
Gardeau Flags and Shales
Nunda Sandstone
Wiscoy Flags and Shales

Chapter 4. Quaternary Geology 56
Glacial Geology
Terminal or Olean-Salamanea moraine
Binghamton moraine
Valley Heads moraine
Hamburg-Batavia-Victor moraine
Medon-Waterloo-Auburn moraine
Rochester-Albion moraine
Oswego moraine

Glacial Lake Development in Western New York
Lake Newberry
Lake Hall
Lake Warren I
Lake Warren II
Lake Dana
Lake Scottsville
Lake Dawson
Lake Iroquois
Gilbert Gulf
Mendon Ponds County Park
The Pinnacle Hills
Brighton-Cobbs Hill
Pinnacle Hill
Divinity-Highland-Mount Hope-Oak Hill
Water Laid Deposits
Till

Glacial Striae

The two lobe or interlobate hypothesis

Chapter 5. Letchworth State Park Geology 70
Geological Features
Lower Falls
Middle Falls
Upper Falls

References 79

Chapter 1. Introduction

Jones and Jones (no published date) provided an introduction to Letchworth State Park in an article posted on the internet from the St. Lawrence Geology Club.

Letchworth Gorge, also called "the Grand Canyon of the East" is a prime example of the mighty Genessee River cutting into thick piles of sedimentary rock. The gorge is approximately 22 miles long and up to 330 feet deep. Three major water falls (cataracts) are present within the gorge. The walls of the canyon represent the past 400 million years of the region's geologic history. The fossil record for the Silurian and Devonian in the area are the most diversified when compared to other regions of the world.

The Genessee River is the only river to completely cross New York State. Its current northward flow has been consistent since the Ice Age. As a result of this, a stratigraphic unit is exposed which reflects the depositional environment and marine life which inhabited this area throughout geologic history.

The rocks observed in the gorge are sedimentary in origin. Sediments represent depositional sediments of the Ordovician, Silurian, and Devonian Periods. On top of the bedrock rests poorly consolidated rocks and glacial deposits belonging to the Pleistocene.

During the Ordovician Period, the region was flooded by a shallow inland sea. As a result, a great deal of sand and carbonate mud were deposited which later compacted into sandstone, shale, and limestone. Also occurring at the same time was the beginning stages of the Taconic orogeny. As the mountains rose in the east, the sediments were eroded and deposited in the west. These sediments eventually formed the Queenston Delta. In the gorge, these sediments occur as the Queenston Formation. This formation consists of red shales and siltstones absent of fossils (**Figures 1 & 2**).

The end of the Taconic Uplift and the beginning of the Silurian Period can be observed in the Medina Group. These sandstones are known for oil and gas reservoirs. Directly overlying the Queenston Formation is the Grimsby Sandstone, representing the dying stages of the Taconic Uplift. The Grimsby is very similar to the Queenston, lacking fossils. The red color in these rocks is the result of iron oxide resulting from deposition in a highly oxidizing environment. The contact between the two formations in uncertain (**Figure 3**).

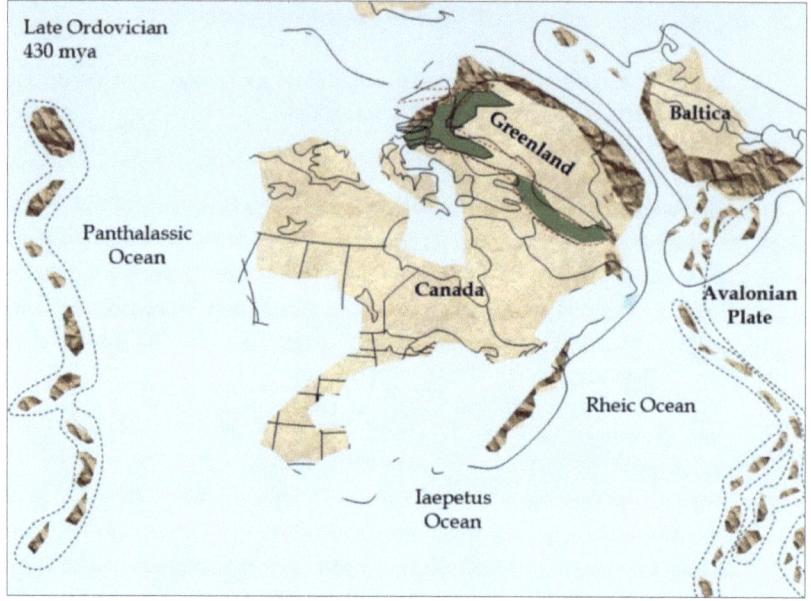

Figure 1. The Rheic Ocean occupied the east coast, covering New York State. The Iaepetus Ocean covered the southeastern part of the U.S. during the Late Ordovician Period 435 million years ago. The Taconic Mountains were beginning to uplift in response to the Avalonia Plate advancing toward the east coast.

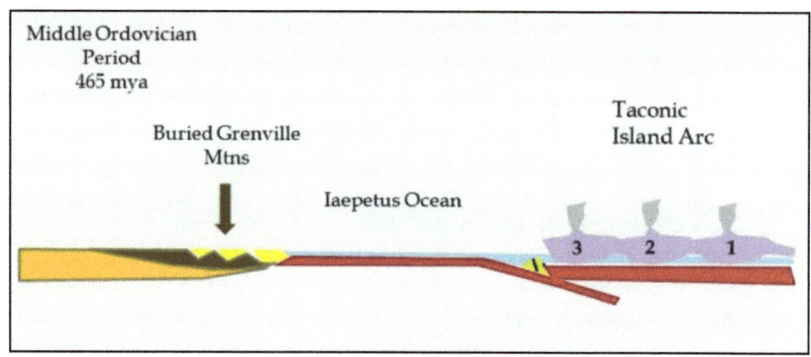

Figure 2. Cross section diagram depicting the tectonics involved with the Taconic Island Arc approaching the eastern U.S. during closing of the Iaepetus Ocean prior to uplift of the Taconic Mountains during the Middle Ordovician Period, 465 million years ago. Numbering scheme represents the progressive positioning of the arc from 1 (earliest) to 3 (latest).

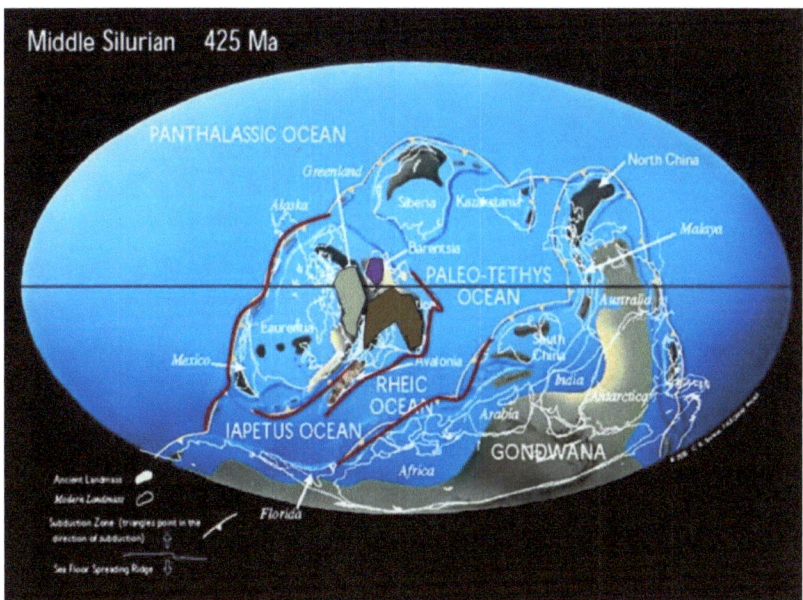

Figure 3. During the Middle Silurian, the Taconic Mountains eroded down to a plain indicated by the tan strip occurring along the eastern U.S. continental margin. Greenland, Australia, and Baltica were amalgamated at the northeastern coast of Canada at this time. New York was positioned about 30 degrees south of the Equator at this time. Avalonia continued to approach to east coast along a subduction zone that existed to the west of the Avalonia Plate. Source of the Paleogeographic maps by C.R. Scotese, 2013 PALEOMAP Project, www.scotese.com.

The Lower and Middle Silurian Period is represented by the Clinton Group. These sediments consisted of shale and thin limestone that represented a quiet time in the absence of mountain building. The Clinton Group contrasts with the Medina Group in the fact that even though a shallow sea environment continued to persist, life began to flourish at this time. The Rochester Shale at the top of the Clinton Group is very fossiliferous. The fossils represented brachiopods, bryozoans, trilobites, and ostracods. The limestone interbedded with this group formed in clear warm water where reefs flourished.

Above the Clinton Group rests the Lockport Group consisting of gray, coarse textured massive dolomite. The dolomite is highly resistant to erosion, forming the cap rock at the falls of Rochester and Niagara. In other regions, the Lockport Group forms massive coral reefs with abundant fossils, representing a warm shallow sea. The upper part of the Lockport Group contains diverse mineral assemblages that formed in solution cavities including variously colored fluorite cubes, clear gypsum, red brown sphalerite, celestite, dolomite, and calcite.

The Late Silurian is represented by the Salina Group consisting of thick deposits of shale and dolostone interbedded with salt and gypsum. These sediments represent a depositional environment that was generally arid in climate and with a shoreline that was composed of shallow bays and lagoons. This type of environment was inhospitable to marine life where very few fossils are present.

The Devonian Period was witness to strong periods of erosion which removed most of the sedimentary rocks. During the Devonian, New York was subjected to additional uplift when Gondwana began to approach the eastern continental margin of Laurentia, closing the Iaepetus Ocean into a narrow shallow seaway. The western part of New York remained submerged, accumulating sediments shed from the rising mountains to the east (**Figure 4**).

The Onondaga Formation developed in the Middle Devonian Period. Limestone formed at the base of the Catskill Delta in a clear, warm, shallow sea where sea life flourished. These sediments are represented by the Onondaga Escarpment located in Syracuse.

Overlying the Onondega Formation are black to blue gray shales with thin interbedded limestone which forms the Hamilton Group. These sediments represent the far edges of the Catskill Delta.

The lower unit shales represent a depositional environment that had a stagnant, poorly oxygenated waters containing sparse fauna. The upper Hamilton Group contrasts with the lower part consisting of calcareous and richly fossiliferous corals, bryozoans, brachiopods, trilobites, crinoids, ostracods, and abundant plant matter.

The Sonyea and West Falls Groups overlie the Hamilton Group. Sediments are black shale at the base grading upward into interbedded shales, siltstones, and sandstones at the top. This sequence represents a prograding delta.

Figure 4. In Late Devonian time, 350 million years ago, Gondwana was closing the Iaepetus Ocean as it approached the eastern continental margin of Laurentia. Arcadian Mountain uplift continued over much of New York but a small portion of the western part of the state remained submerged, collecting sediments shed from the uplifting region to the east.

Chapter 2. General Geology

The University of Rochester (1956) posted a guidebook on general geomorphology and stratigraphy presented in the next two chapters.

Geomorphology

The area of study has maximum dimensions of about 60 miles east-west and 50 miles north-south. It reaches from the southern shore of Lake Ontario across a portion of the Central Lowlands physiographic province into the highlands of the Glaciated Allegheny Plateau section of the Appalachian Plateaus province. Throughout the area geologic phenomena recorded in observable features represent two widely separated phases of development: (1) the bedrock geology dating to the Middle Paleozoic and (2) the period of Pleistocene glaciation and postglacial developments. The gap between these two is bridged only by the obscure and not too spectacular tectonic development of the area and by the more easily interpreted stages in the evolution of the surface of the land itself.

The area between Lake Ontario and the Allegheny Plateau is the far eastern attenuation of the great Central Lowlands province, which, as a whole, includes an area of over half a million square miles, comprising a part, or the entire portion, of the land within sixteen states. Within this large area the variation in essential morphologic form is no greater than in the smaller physiographic provinces, lack of major distinctions being partly a result of the nearly flat-lying strata. The most dominating characteristic in the local area is the effect produced by glaciation.

Similarly, if it were not for the glaciation in the northern part of the Allegheny Plateau section, the glaciated and unglaciated portions of that section would probably not be considered distinct, since the extent of dissection and the amount of relief of both areas are comparable. Throughout the area, bedrock consists of Middle Paleozoic sedimentary formations virtually undeformed except for imposition of a regional southward dip which averages about 60 feet per mile, with local variations. The land surface slopes gently toward the north, declining from a hilltop level of nearly 1600 feet at the southern limit of the area to 246 feet at the shoreline of Lake Ontario & Outcrop belts are thus east-west in extent across the area and those of the more resistant formations are marked by low, but in places abrupt escarpments.

Largest of these are the Onondaga escarpment (generally taken as the dividing line between the Central Lowlands and the Allegheny Plateau), which extends with slight interruptions nearly from Lake Erie to the Hudson River, and the Niagara cuesta (on the Lockport dolomite), which is best developed from somewhat west of Rochester westward to the Niagara River. Between the two lie the thick and weak Salina shales of the Upper Silurian.

South of the Onondaga escarpment the land rises at first gradually over the calcareous shales of the Middle Devonian and then more rapidly over the more resistant shales and sandstones of the Upper Devonian. Although none of these outcrops is highly resistant, the difference between firm sandstones and weak shale is reflected in places by a noticeable change in altitude and style of surface. In the vicinity of the boundary between the two provinces, dips of beds are a bit steeper than average and the northward slope of the surface is pronounced. Therefore, the outcrops of the formations near the boundary are narrow in comparison to outcrops of formations of equal thickness farther from the boundary.

Physiographic History

Near the end of the Devonian period the western part of New York was lifted out of the Paleozoic sea. Evidences of the drainage systems which followed, throughout the remainder of the Paleozoic and the entire Mesozoic eras, have been much obscured by subsequent Tertiary erosion; however, the earliest drainage was undoubtedly consequent and paralleled the south slope of the land. It is believed that since Devonian time the streams have removed at least 2000 feet of bedrock above Rochester and vicinity.

Since the most primitive drainage system was first established, across New York from Canada, streams have been affected by at least two periods of great continental uplift and one epoch of continental glaciation. The first great uplift was caused by the Appalachian Revolution in Permian time which, without much crushing or folding in this area, raised and created for the first time a broad and high Allegheny Plateau. The crustal folds of northern Pennsylvania were continued into New York as very weak anticlines and synclines but the elevation of the Plateau in New York was probably sufficient to keep the enlivened or rejuvenated streams in their southward courses.

Succeeding the Appalachian Revolution came the long era of the Mesozoic during which New York appears to have experienced a period of relative rest from extensive epeirogenic movements. During this era western New York was subjected to continuous erosion and ultimately transformed to a low-altitude peneplane, as evidenced by the near-constant elevation of hill tops in southwestern New York.

Probably the streams were mostly held in their old courses in the lowering land surface. The oldest valleys in central and western New York indicate a direction of stream flow to the south and southwest, the very thick and weak shales of the Ordovician and Silurian, along the belt of the present Mohawk and Ontario valleys, encouraged the production of subsequent valleys by the deepening of east-west streams tributary to the great south-flowing trunk streams. It is possible that these dominating valleys were initiated during Cretaceous time, but their great development dates from Tertiary time.

Following the long standstill with erosion and base leveling of Cretaceous time came the second long epoch of continental uplift in the Tertiary. It is probable that the great rise in the Tertiary was not a single lift but intermittently gradual and oscillatory with possibly long erosional intervals. With the uplift, the streams revived and deeply entrenched their valleys in the elevated Allegheny Plateau, producing high relief and anomalous drainage. This enlivening of the drainage in early Tertiary time gave expression to the different resistances of the rock strata.

Previously, the direction of stream flow seems to have been governed primarily by the south slope of the land. At this time, however, the resistance to south-flowing waters affected by east-west outcrops of competent beds such as the Lockport or Onondaga limestones gave rise to a gradually expanding subsequent stream pattern. Tributary and subsequent streams flowing east-west which happened to be fortunately situated on weak rocks, deepened their channels more rapidly than even large south-flowing trunk streams. The Ontario Basin was cut by such a subsequent stream into the soft Upper Ordovician Queenston shale. This stream, the mighty Ontarian River, found an outlet to the sea either by way of the Mississippi River or the St. Lawrence depression. Its tributaries from the south captured great amounts of water from the old south flowing system, making possible rapid and extensive deepening of some of the old river courses (**Figure 5**).

In tracing the development of present day topography, it is necessary to consider the effects of Pleistocene glaciation. The Tertiary drainage system was completely disrupted by the sea of ice which, at the time of its greatest extension, had moved in a great mass southward over New York state and into Pennsylvania. There are no continuous moraines near the limit of the ice advance along the New York-Pennsylvania border and scarcely even glacial debris on the uplands, although valleys seem to be well supplied with outwash material.

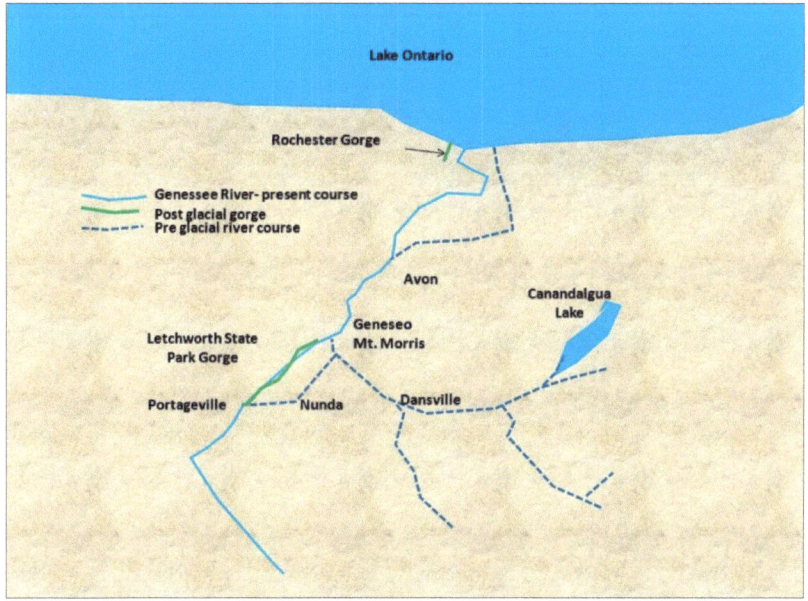

Figure 5. Pre-glacial and present day courses of the Genessee River.

It was not until the ice front had retreated almost to the southern ends of the present Finger Lakes that a long halt occurred and substantial moraines were built. These are known collectively as the Valley Heads Moraine, and it marks approximately the present divide between the Ontario and the Susquehanna drainage except for the Genesee River. North of this line of moraines the time of ice occupation was much longer than south of it and the ground moraine becomes a more important topographic feature.

The constructional effects of the glacial ice are evident in the variety of morainal deposits which abound through the area. Opinions differ, however, as to the erosive power of the ice sheet. The glacier did not have great eroding powers, but that there was a "sort of sandpapering of the land surface" with the greatest work being that of transporting weathered debris was the current theory.

Others, especially more recent workers, felt that the erosional effects of the glacier are very striking, attributing to this many features of the Finger Lakes region: hanging valleys, steepened lower valley slopes, smooth and straight valley walls, and the absence of projecting spurs.

History of the Genesee River

The oldest of the stream valleys tributary to the Genesee were originally a part of the pre-Tertiary southward drainage discussed in the preceding section. The inception of the Genesee as a north-flowing stream in the midst of the early Tertiary south-flowing systems was brought about by the development of the Ontarian River which was largely responsible for the pre-glacial excavation of the Ontario Basin. The original mouth of the Genesee, at its junction with the Ontario Basin, was probably somewhere north of what is now Irondequoit Bay. Aided by a steep northward gradient into the Ontarian Valley, the Genesee actively extended itself southward through head ward erosion and, in so doing, captured the waters from the older south-flowing system.

The ancient Genesee had two main branches. The western branch followed the course of the present river south of Portageville. The eastern and larger branch, called the Dansville branch, was entirely extinguished by the glacier. It appears to have derived its main source of water from the valley now occupied by Canandaigua Lake, and its flow carved a wide valley through the sites of the present towns of Naples, North Cohocton, Wayland, and Dansville. Before the ice age, the western branch probably flowed northeast from Portageville through the valley at Nunda and joined the Dansville branch somewhere in the neighborhood of Sonyea. The united branches continued northward as the main trunk to about five miles north of Avon where it turned east along the outcrop of the soft Salina shales for some 13 miles to the vicinity of Fishers. From Fishers the river flowed northward through the deep Irondequoit Valley.

Judging from the present form of the eastern and western branches, the Dansville one must have carried the principal stream and the western one a tributary.

The advance of the Pleistocene glacier obliterated the drainage pattern. Upon recession of the ice, stream channels had become so disrupted by fillings of glacial debris that important changes resulted in the course of the Genesee as it began again to flow through the area. One effect was the extinction of the entire Dansville branch from Naples to Sonyea.

There may be well over a thousand feet of glacial debris in most of the old river bed east of Dansville and as much as 600 feet north of Dansville to Lake Ontario. Three miles north of Dansville a drill penetrated 450 feet of glacial material without striking bedrock.

Another effect was the blocking of the course of the western branch through the old Nunda valley. This was accomplished by the Portageville moraine, a local development of the Valley Heads Moraine mentioned above. Downstream from Portageville, the western branch was forced to cut a new course until it plunged down the western slope of the old pre-glacial valley near the present town of Mt. Morris (**Figure 6**).

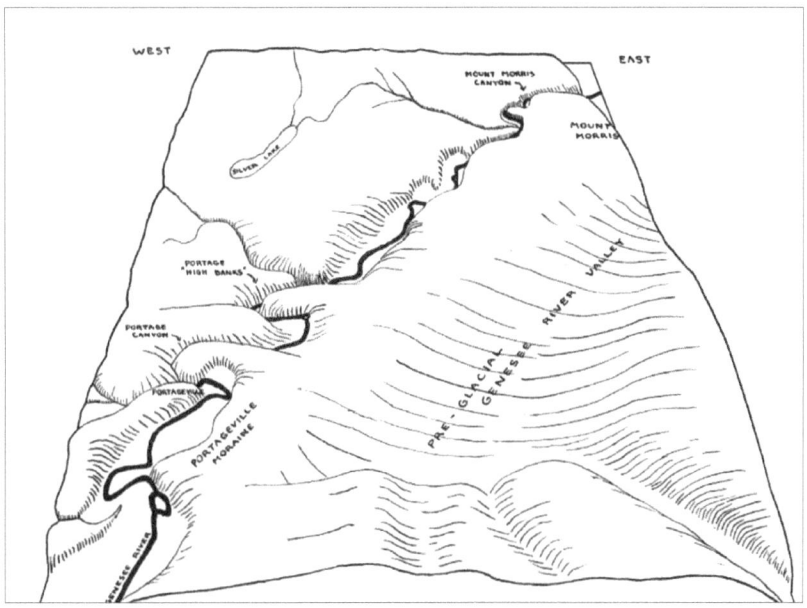

Figure 6. The Genesee River from Portageville to Mount Morris and the glacial landscape that affected its course. Source: The University of Rochester 1956.

This 20-mile post-glacial portion of the Genesee lies mostly within the limits of Letchworth State Park and comprises a gorge that has been aptly termed the "Grand Canyon of the East".

From Mt. Morris, the Genesee flows in its old valley northward to a point beyond Avon where it had previously swung east along the belt of soft Upper Silurian shales. Here another effect of morainal filling is evidenced. The river was blocked from its former course and flowed directly northward to Lake Ontario. The upstream portion of this stretch is a gently meandering passage across a surface with low gradient, composed of ground moraine and a veneer of glacial lake sediments.

The portion of the course now running through the city of Rochester is another gorge cut deeply into bedrock.

Upper Gorge -- Portageville to Mt. Morris

At Portageville, there is a striking contrast between the physiographically old valley of the Genesee and the very young valley which cuts the Portage Canyon. From a broad open valley with cultivated slopes, the river swings suddenly into a narrow steep-walled chasm. Within 3 miles after entering the Portage Canyon, the water drops a total of 317 feet (from 1077 feet to 760 feet above sea level) over three major falls and a series of rapids. The upper and lower falls produce a drop of about 70 feet each while the middle falls drops 107 feet. Fall maker for the upper falls is the Nunda sandstone, the middle and lower falls are over resistant sandstones in the Gardeau formation. All rocks exposed in the canyon are Upper Devonian in age. For a mile and a half beyond the Portage Canyon, the river flows in an open valley from which it plunges into another deep steep-sided ravine which is called the Portage High Banks. In dimensions, this ravine is superior to the Portage Canyon upstream, being 500 feet in depth. It has no cataracts, although the steep rocky slope of the river bed produces a swift tumbling current. This ravine extends for about three miles. The valley then opens again for about a 5 mile stretch and finally narrows into a third gorge, the Mt. Morris Canyon through which the river flows 7 miles before returning to its pre-glacial valley.

Near its north end, the Mt. Morris Canyon is spanned by a large flood control dam. Leaving the Mt. Morris Canyon, the river flows again into its pre-glacial valley. From this point the abandoned Dansville branch of the valley extends off to the southeast.

Mt. Morris to Irondequoit Bay

The pre-glacial course of the Genesee from Mt. Morris north to Avon is a broad, flat bottomed valley with high, but rounded slopes to the sides. The valley was probably deepened by action of glacial ice; beneath the valley floor today, are thin flood-plain deposits and glacial lake beds, and then a filling of glacier-transported rock debris to an unknown depth (at least several hundred feet). Along the valley floor the river flows sluggishly through elaborate meanders.

The preglacial channel eastward from Avon to Fishers and thence north to Irondequoit Bay is nowhere evidenced on the surface south of Irondequoit Valley itself. The channel was entirely filled in with glacial debris but its course has been established with reasonable certainty from records of water wells.

This old filled valley, which continued northward out into Lake Ontario beneath the valley re-excavated by Irondequoit Creek and bearing Irondequoit Bay, carries a strong flow of subsurface water which has been important in supplying both municipal and industrial developments in the area immediately east of the city of Rochester.

Avon to Rochester

The portion of the post-glacial course from Avon to Rochester is transitional in aspect between the maturity of the old valley to the south and the obvious youth of the Rochester gorge. The low gradient of the postglacial surface on which the river flows here is responsible for an apparent physiographic age in excess of that indicated by other criteria.

In Rochester the river descends over three major falls dropping a total of 235 feet to the level of Lake Ontario. The fall makers and drops at the three falls are: upper (Lockport dolomite), 90 feet; middle (Reynales limestone), 42 feet; lower (Thorold sandstone), 97 feet. Exposures in the gorge walls constitute one of the classic sections of sedimentary rocks in the country, extending from the Queenston shales at the base up through the Lower and Middle Silurian to the Lockport dolomite. Unfortunately access to much of the section is difficult, but new exposures from the Queenston up into the Clinton group occur in Durand-Eastman Park, Rochester.

Chapter 3. Stratigraphic Succession

Ordovician System

Upper Ordovician

Queenston Shale

The oldest formation exposed in the Rochester area is the Queenston shale of Upper Ordovician (Cincinnatian) age. The Queenston is about 1000 feet thick and underlies much of the Ontario basin to the north; only the upper 55 feet are exposed in the Genesee Gorge.

Lithology

The Queenston shale in western New York is made up predominantly of thin-bedded, earthy red shale, breaking up readily into angular fragments and weathering finally to red clay. At some horizons, notably near the top of the formation, heavy red sandstone layers occur, varying from a few inches to a few feet in thickness, often resistant, but weathering into shaley fragments. The red color is caused by thin coatings of ferric oxide on detrital grains. Where iron oxide is absent, or ferrous, rather than ferric iron prevails, gray, green, or white blotches or layers result. Among the red shale layers are thin bands of green shale, usually one or two inches in thickness of the same appearance, aside from color, as the red shale in which they occur. These bands are fairly continuous within a single outcrop and may extend for many feet, but cannot be traced over great distances.

The texture of the shale is rather fine and earthy, some of the particles, however, being visible to the naked eye. The sandstone layers are also fine grained, being clearly distinguishable from the typical Grimsby sandstone, which, although almost identical to the Queenston in color, is coarser, frequently quartzite, and sometimes politic in appearance.

Depositional Environment and Regional Relations

The Queenston of New York is correlated with the Juniata of the Appalachian area. Both represent deposits laid down on partly subaerial deltas built up by streams eroding highlands to the east, which were elevated toward the end of Ordovician time. To the west this red bed facies graded into normal marine facies now represented by the fossiliferous Richmond strata exposed in the Cincinnati area in Ohio-Indiana-Kentucky.

The placement of the contact between the Ordovician and Silurian systems has been widely discussed and hotly debated in stratigraphic literature for many years by some of America's foremost geologists. Specifically the problem in eastern North America is to determine whether the beds of Richmond age belong to the Silurian or to the Ordovician. The temper of current times favors placing the Richmond in the Ordovician.

In the Niagara gorge section, the red Queenston is overlain by the white Whirlpool sandstone. There the marked lithologic contrast between the two provides a readily recognized horizon at which to locate the systemic boundary. In the Rochester gorge, the red Queenston is overlain by equally red Grimsby sandstone and the boundary has been more arbitrarily determined to lie at the base of the first heavy-bedded siltstone layer. From what can be seen in the field and under the microscope, sedimentation here was fairly continuous across the boundary.

Silurian System

Lower Silurian

Grimsby Sandstone

Lithology and fossils

The Grimsby sandstone (55 feet thick in the Rochester gorge) is made up mostly of heavy-bedded red siltstones but contains, in addition, heavy-bedded gray siltstones, red graywackes, thin-bedded to fissile, red argillaceous shaley partings, and thin, fissile, green argillaceous rocks. In terms of total composition, the Grimsby has slightly less clay and more quartz than the Queenston. The Lower Silurian rocks of westernmost New York reflect two major facies: a red continental facies, the Grimsby, to the east, and a marine facies to the west. Only the red siltstone facies is represented at Rochester.

The Grimsby is dominantly red in color, but especially in the upper 20 feet, are found spots and blotches and even thin layers of a peculiar light green color similar to those in the Queenston. In some places the green blotches seem to follow the bedding very closely, but in the majority of cases they are irregular, crossing bedding planes, and passing from one layer to another. As a whole the green blotches seem to be confined to the more sandy layers. Cross-bedding, which is prominent in the red sandstone, is either very indistinct or loses its identity entirely in the green blotches.

Fossils are scarce in the Grimsby. Arthrophycus alleghaniensis and Daedalus archimedes, variously attributed to worms or seaweeds, are the only common evidence of past life.

Substantial evidence has been offered to show that the Grimsby is of beach origin, at least the part which outcrops at Rochester. Such evidence as ripple marks and mud cracks are plentiful. The cross-bedding shows no apparent orientation and several different types can be observed. Furthermore, the sediments composing the Grimsby are not uniform in size or weight. Small pieces of shale are found even in the conglomeratic layers. The deposit is poorly sorted. Several layers of arkosic conglomerate occur in the section, especially near the top. The sand grains and pebbles which comprise the major part of the unit show a variety of shapes: the constituents in some layers are rounded and spherical, in others, angular and non-spherical, and in still others there is a mingling of the two types.

Correlatives east and west

In the Niagara gorge, the basal Silurian white Whirlpool sandstone, already mentioned, is separated from the red Grimsby by an eastwardly extending tongue of the marine sequence which is still better developed farther west in Ontario. In southeastern New York and in eastern Pennsylvania the Grimsby is represented by the Shawangunk conglomerate, and by the Tuscarora in central Pennsylvania and through the Appalachians southward to Tennessee (**Figure 7**).

Middle Silurian

Clinton Group

Thorold Sandstone

Lithology and age

The Thorold sandstone in the Genesee Gorge is a light gray-green, fine-grained resistant siltstone, with a maximum thickness of 5 feet. Its resistant character makes it stand out; it is the fall maker for the lower falls of the Genesee. In color it's an easily recognizable unit, being referred to locally as the "gray band". Through western New York the formation retains its dense, compact character which, together with its light gray color, sets it off from the underlying Grimsby with which it is gradational.

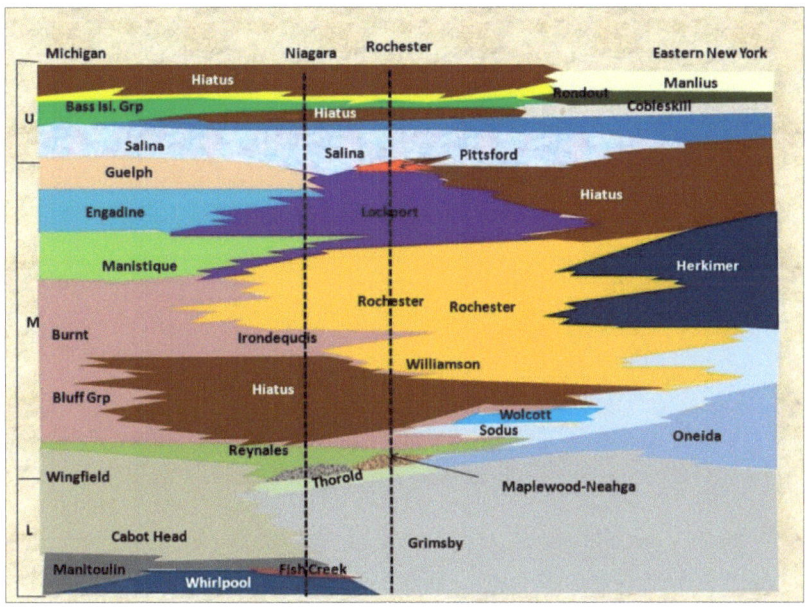

Figure 7. Generalized diagram suggesting relationships of major Silurian rock units east and west of the Rochester area.

Argillaceous material is abundant in the form of thin shale partings. The sand grains range from angular to semi-angular. The quartz of the Thorold is on the whole slightly smaller in grain size and less rounded than that in the underlying Grimsby. The cementing material is both siliceous and calcareous, with silica having the dominant role. The Thorold contains many thin shale breaks similar to the shale of the overlying Maplewood, but with a higher percentage of quartz.

A red band or lens a tenth of a foot thick occurs slightly over a foot above the base of the Thorold in the Genesee gorge. This has proved to be a Clinton type "iron ore" more siliceous than the younger Furnaceville, but essentially a limestone which has experienced considerable replacement by hematite. As there is no characteristic typical of the Clinton as the "iron ores", this ties the Thorold, at Rochester, to the Clinton group. It has been suggested that the Thorold represents reworked Grimsby. At any rate, it surely marks the beginning of a re-advance of marine conditions from west to east over the greater Queenston delta. It is, therefore, a transgressive unit, and, along with some of the succeeding thin stratigraphic units of the lower Clinton, decreases slightly in age from west to east. It is properly a part of the lower Silurian in the west and of the Middle Silurian in the east.

Fossils

For the most part the Thorold contains very few fossils. Arthrophycus alleghaniensis has been found, and a small form of Daedalus.

Maplewood Shale

Lithology and origin

The Maplewood shale (21 feet thick in the Genesee gorge) gradationally overlies the Thorold. It is a smooth, slightly calcareous, green, platy shale, the lower 3 feet being sandy and much more calcareous than the rest. At the lower contact, silt to very fine sand makes up over 50 percent of the matrix. An abundance of phosphatic nodules is characteristic of this lower Maplewood in Monroe County. Several thin limestone beds, all under one inch in thickness, occur in the Maplewood.

They are fairly continuous for many feet within a single outcrop, but it is doubtful if they are reliable horizon markers over long distances. Individual bedding planes of the shale generally cannot be traced, due to jointing and fracturing, and the intense hackly weathering. Interesting swirls of uncertain origin can be seen on some of the bedding planes.

The upper portion of the Maplewood contains interesting flattened calcareous pellets, many of them about the size and thickness of a dime. These may be either clay pellets with a coating of calcite acquired by being rolled around in a lime mud and flattened by compaction, or the result of deposition of the calcite by solution waters moving along joints and fractures which occur in abundance in the formation. There appears to be a concentration of these calcite spots along joint planes.

The origin of the rather homogeneous, fine green Maplewood is not clear. It has been suggested that it represents a winnowing of the red Grimsby, being the finer material, whereas the Thorold is the coarse material left behind.

Extent and fossils

From its known maximum thickness of 21 feet in the Genesee gorge, the Maplewood thins to the east and west. It is correlative with the lithologically similar Neahga shale of the Niagara section and may be continuous with it. Fossils are rare in the Maplewood, with most of the reported forms coming from the lower 3 feet. A microfauna was described from the Maplewood.

The Reynales limestone (17 feet) in the Genesee gorge consists of three members: the Brewer Dock member (3 feet), the Furnaceville iron ore or hematitic limestone (8 inches to 2 feet), and the upper Reynales limestone (13 feet).

The Brewer Dock is a local member in the Genesee gorge, consisting of 3 feet of gray calcareous beds with shaley partings, lying between the green Maplewood and the red Furnaceville. Heavy mineral studies indicate that the dark argillaceous layers are very like those of the Maplewood, differing principally in color and a greater carbonate content. The Brewer Dock is gray when fresh and buff when unweathered. The crystalline, medium gray limestone is like that of the upper Reynales. The characteristic fossil is a minute gastropod of the genus Cyclora; the brachiopod Hyattidina congesta is also found. Small phosphate pebbles also occur in the formation, particularly in the basal limestone bed; some are fossil molds.

The Furnaceville iron ore is one of the many lens like, hematite-rich layers which characterize the Clinton group. This bed thins to the east and is absent a few miles from the Rochester gorge at Irondequoit Bay. A few miles still farther to the east, the Furnaceville is again found, but here it occupies a position at the base of the Reynales. The unit can be traced as far west as western Monroe County.

The Furnaceville is a thin, highly variable, hematitic limestone. The variation in hematite content results largely from the presence of thin shale breaks and layers of non-hematitic limestone. The Furnaceville is dominantly a fossiliferous ore, the hematite having replaced brachiopods, bryozoans, crinoids, ostracods, etc. Both contacts of the Furnaceville are sharp, although there are indications that there was a transition from the hematite-forming conditions to those of the typical Reynales.

Above the Furnaceville in the Genesee gorge are over 13 feet containing layers replete with shells of the brachiopod Pentamerus, interbedded with crystalline dolomitic limestone. Some of the crystalline limestones are fossiliferous, but most of them are barren. A few argillaceous unfossiliferous limestone layers are also present. Thin shale partings are found throughout, but are concentrated in the lower two-thirds of the unit. Cherty beds occur in the same portion. The beds of the upper Reynales are thicker, with less shale toward the top of the formation. Unusual cross-bedding occurs in the limestone layers in the upper 7 feet.

Westward the Reynales is continuous into Ontario and probably has correlatives in Michigan. To the east it grades into shale, and finally it and the overlying Sodus shale become inseparable.

Rochester Shale

Lithology

The Rochester at its type locality in the Genesee gorge is about 85 feet thick. The formation is considered to be the highest member of the Clinton group. Except for the basal 10 feet, which is brownish gray, it is dark bluish gray in color. In all but these lower 10 feet limestone layers are plentiful. The lower 25 to 30 feet of the formation is a weak shale which, upon being exposed, quickly disintegrates into a blue to brown clay. The upper 20 to 25 feet is more massively bedded and slightly dolomitic. To the east it thickens and becomes more clastic and is equivalent to the Herkimer sandstone in large part.

Upper contact

The relationship between the Rochester and the Lockport has been subject to dispute. In some places there is a decidedly sharp lithologic change while in others there is an interbedded gradation. Some authors, on the basis of faunal correlations with the mid-continent area, have suggested a considerable break between the two. Others have postulated continuous deposition. In the Rochester area, the Rochester grades into the Lockport without sign of a physical break of any consequence.

Fossils

The lower few feet and the upper 15 feet are relatively unfossiliferous, but the rest contains an abundance of fossils. The Rochester shale contains by far the most varied fauna of the Clinton group in this area. Brachiopods form the dominant element. Some of the more common forms are: Parmorthis elegantula, Sowerbyella transversal, Stropheodonta profunda, Schuchertella tenuis, Atrypa "reticularis", Leptaena rhomboidalis, Whitfieldella and Camarotoechia neglecta, Rhipidomella Qlbrida, and Dictyonella.

Bryozoans are also common, with Mesotrypa, Ceramopora imbricatall and Chasmatopora asperatostriata most abundant. The cephalopod Dawsonoceras annulatum, and the trilobites Dalmanites limulurus, Trimerus delphinocephalus, and Arctinurus also occur in appreciable numbers.

Lockport-Guelph Group

Lockport Dolomite

Lithology

The Lockport is a sugary, gray, massive dolomite, in places sandy. The sugariness in some places is due to the coarse dolomite grains, although to the east it actually becomes quite sandy. The formation commonly contains small cavities lined with dolomite and other crystals. In outcrop it ranges up to 150 feet thick, while in wells to the south it is reported to be about 300 feet thick. Its lithology is distinctive and the formation is a resistant unit, forming the crests of Niagara Falls and the upper falls of the Rochester gorge. Between these two areas it is responsible for the Niagara cuesta. The Lockport has been traced as far east as Utica.

Guelph facies

West of Rochester into Ontario and the Michigan Basin, the upper part of the typical Lockport gives way to a different facies, the Guelph dolomite which has one of the most striking lithologies in eastern North America. It is a sugary dolomite of a very light tan to almost white color. It is very fossiliferous in places, with an unusual fauna. The Guelph facies is not clearly represented in the Rochester area.

Fossils

Two faunas are present in the Lockport dolomite in western New York. The lower or normal Lockport fauna, although not abundant is derived from the underlying Rochester shales, and is found best developed in the more argillaceous portions, while the crystalline limestone portions contain numerous crinoid fragments. The upper, or Guelph fauna, is derived from Canada to the west. Some of the faunal elements of the Guelph are present near the top of the formation, although the lithofacies is not of typical Guelph aspect, but rather resembles the rest of the Lockport. Fossil collecting is not generally good in the Lockport.

Upper Silurian

The stratigraphic terminology applied to the Upper Silurian has had a long and complex history. The scheme followed here differs slightly from the terminology of the Silurian correlation chart. The Salina is considered a formation with two major facies: lying above the Pittsford (a local formation) or the Lockport. The Salina is, in turn, overlain by the Bertie formation which is thus excluded from the Salina.

Salina Formation

Lithology

The 600-900 feet of sediments which lie between the Pittsford (or the Lockport where the Pittsford is absent) and the overlying Bertie formation consists of a variety of lithologic types which customarily have been allotted two or three formation names: Vernon shale, Camillus shale, and sometimes Syracuse salt.

Two major facies are represented, a red and green argillaceous facies, and a gray to brown more calcareous facies with evaporites. These have a complex interrelationship which is more realistically reflected by using the names Vernon and Camillus for the red and gray facies respectively, than by recognizing these two as distinct formations, the Vernon below the Camillus, even though most of the red beds occur in the lower part of the sequence. The salt of the Upper Silurian is too universally associated with what is here called the Camillus facies to warrant recognition as a separate unit.

Salt, gypsum or anhydrite account for up to 300 feet in some subsurface sections; all of these evaporites have been removed by leaching from surface exposures. Both the Camillus facies and the overlying Bertie formation thin to the east and disappear along the Mohawk Valley, where they change facies and are represented by the Brayman shale.

Fossils

The fauna of the Vernon facies is even more impoverished than that of the underlying Pittsford. The eurypterids Fterygotus vernonensis and Hughmilleria phelpsae are characteristic, but not common. The Camillus facies is nearly barren; almost no megafossils, other than rare specimens of Ctenodonta salinensis appear to be present.

Bertie Formation

Lithology

The Bertie formation is 50-60 feet thick in this region and is composed of drab or gray limestone, dolomitic limestone, and dolomite. In western New York, the formation has been divided into four members which are in ascending order: Oatka shale, Falkirk dolomite, Scajaquada shale and dolomite, and Williamsville dolomite. The identity of the basal member is poorly established.

The Falkirk consists of massive beds of dark dolomite weathering yellow-brown and characterized by a large-scale conchoidal fracture. The Scajaquada consists of medium-bedded dark dolomite with considerable argillaceous content. The Williamsville is a dark brownish-gray dolomite with pronounced conchoidal fracture, the rock is laminated and weathers light gray. The upper two members are exposed at Oaks Corners.

Fossils

The Bertie is noted for its eurypterid fauna, of which the more common forms are: Eurypterus lacustris, remipes, Eusarcus scorpionis, and Pterygotus buffaloensis. Fragments of these fossils are occasionally found, but specimens anywhere nearly approaching whole individuals are exceedingly rare.

Cobleskill Formation (Akron dolomite facies)

Lithology

The Cobleskill formation as defined herein as consisting of two facies: an eastern fossiliferous limestone facies and a western dolomitic facies known as the Akron dolomite. The change from the limestone facies to the dolomite facies is not abrupt. At Oaks Corners, the Cobleskill is a fine grained brownish-gray dolomite in massive layers having stylolites, irregular fracture, and occasional geodes. Farther west the dolomite becomes quite mottled, thin bedded, and presents a banded appearance.

The Cobleskill is unconformably overlain by the Onondaga limestone and gradationally underlain by the Bertie formation. The thickness of the Cobleskill ranged from 8 to 20 feet.

Fossils

To the east the limestone facies of the Cobleskill is abundantly fossiliferous. The western dolomite facies has a much more limited fauna which at Oaks Corners is represented solely by silicified stromatoporoids.

Silurian Depositional History

There was no sharp change in conditions at the end of the Ordovician. The Silurian began with a continuation of continental deposition in the east (Taconic disturbance), with a widespread shallow sea over most of the Interior Lowlands. The Lower Silurian is marked by coarse clastics to the east and fine clastics to the west.

Like the Queenston (Ord.), the Medina sandstone represents deltaic deposition in parts and marine in other minor parts. In the east the Lower Silurian is represented by the Shawangunk conglomerate and to the south and east by the Tuscarora sandstone. In the Niagara gorge, the lowermost Silurian is represented by the white Whirlpool sandstone, thought by some to represent a dune sand.

Farther to the west, in Ontario, the Lower Silurian is represented by a marine sequence, the Cabot Head shale, while still farther to the west it grades to the calcareous Manitoulin and Mayville formations. The shoreline was located in extreme western New York, getting as far as Ontario at times.

The same gradation of facies exists in the Middle Silurian, but the facies all shifted markedly to the east. Continental deposits are found only in eastern New York, with marine shale from eastern New York to western New York, while calcareous facies extended as far east as Rochester, and a little beyond at times.

There were also widespread seas in the Central Stable Region during the Middle Silurian. However, there is a contrast with the Lower Silurian in that there was a lack of the tremendous thickness of clastics to the east. The general shift of facies to the east was probably due to the reduction in the supply of clastics while subsidence continued. The facies shifted back and forth, however, with a few tongues of the calcareous facies extending east beyond Rochester (Irondequoit, Reynales), while the clastic tongue extended from the east to the west (Rochester).

In the Niagara gorge, the entire Middle Silurian is represented by about 100 feet; in the Rochester gorge it is 165 feet; it is thickest near Syracuse where it is 315 feet. The difference in thickness is due to the appearance of more clastics to the east, and also to the fact that some of the section is missing in the western part.

The Thorold sandstone probably represents a reworking of the upper layer of the red Grimsby. The Thorold becomes progressively younger to the east, due to a transgression of the sea in that direction. The Maplewood is thought by some to represent a winnowing of the red Medina, being the finer material, while the Thorold represents the coarser material. East of Rochester it disappears, due either to subaerial erosion or exhaustion of clastics.

After deposition of the Maplewood, the depression of the geosyncline continued and marine waters became clear and spread wider, forming the Reynales limestone. The Reynales becomes argillaceous to the east and is lost in the longitude of Syracuse in a shale.

Near the base of the Reynales in the Genesee gorge is the Furnaceville iron ore. There are other iron ores in the Silurian, but this is the most extensive.

A break between the lower Sodus and the Irondequoit in the Rochester area represents a time when this area probably was above sea level, although a basin of deposition existed to the east in which middle Clinton deposits accumulated in central and eastern New York.

Upper Clinton time is indicated in the Rochester area by Williamson and Rochester shales, separated by the Irondequoit limestone which feathers in from the west. All three formations are lost in the east in the Herkimer sandstone. The termination of the Clinton brought about a widespread general submergence which produced conditions necessary for Lockport sedimentation.

In the Upper Silurian, there was a gradual change to hypersaline conditions, resulting in thick deposits of evaporites in Michigan and western New York. The lower Upper Silurian is marked by the greatest restriction of the Paleozoic seas. The areas of deposition were restricted to the Michigan and the western New York, north central Pennsylvania basin. For a large part of the early Upper Silurian, these basins may have been connected. They were slowly depressed and vast thicknesses of fine clastics, dolomites and evaporites accumulated in them.

Silurian-Devonian Hiatus

In the Rochester area the uppermost Silurian and a large portion of the Lower Devonian are represented by a major disconformity between the Cobleskill and the Onondaga formations. At the close of the Silurian, marine waters withdrew from the mid-continent area and were limited to the eastern portion of the Appalachian geosyncline. The uppermost Silurian (Rondout, Manlius) and lowest Devonian (Helderbergian) formations are restricted to eastern New York, forming a wedge of sediments which thins and disappears to the west.

The Oriskany sandstone, representing the Deerparkian stage, is the first Lower Devonian formation which extended much beyond the axial portion of the geosyncline. It is not distinctly represented in outcrops in the Rochester area, but in the subsurface to the south it is fairly extensively developed and is one of the significant gas-bearing horizons.

The Onondaga was the first really widespread Devonian formation, representing the initial phases of the principal marine inundation of the Devonian. The erosion interval beneath the Onondaga had considerable duration in the Rochester area.

It involved not only pre-Onondaga erosion, which brought about partial removal and reworking of the Oriskany sandstone, but also pre-Oriskany erosion which removed parts of the underlying limestones.

Devonian System

Lower or Middle Devonian

Onondaga Limestone

Extent and thickness

In New York the Onondaga extends with remarkable constancy of lithology from Albany to Buffalo and then continues westward through the greater Great Lakes region. In Pennsylvania, Maryland, and northern Virginia it is represented by a calcareous shale. In the Livonia salt shaft, 25 miles south of Rochester, the Onondaga is 140 feet thick but no complete section outcrops in the Rochester area. Extensive quarries however, afford excellent exposures of parts of the formation.

Base of the Onondaga

The unconformity between beds of Onondaga age and older strata has been recognized as continuous over about 700 miles from eastern New York to central Indiana. Westward across New York State the Onondaga rests successively on older formations. In the Albany-Schoharie area and in southeastern New York in general, the contact with underlying rocks is gradational and there is a facies relationship between the Onondaga limestone and the Schoharie grit. The fine grits of Schoharie age appear to pass very gradually into the impure limestone beds at the base of the Onondaga without any indication of a physical break.

In central and western New York and in portions of eastern New York, there is conclusive evidence that the Onondaga was deposited on an old erosion surface which was emergent until shortly before Onondaga time. The physical evidence for the disconformity in this region includes both an irregular contact surface between the Onondaga and underlying beds, and a basal sandy zone or conglomerate which is not present at all in basal exposures of the Onondaga.

Where these basal clastics occur, they are usually less than a foot thick and frequently grade upward into the limestone, gradually merging with it. It was thought at first that this sandy zone represented the Oriskany. It is now thought that this sandy zone, at least in most instances, was formed by the reworking of Oriskany deposits in the early Onondaga sea.

Twenty-five miles south of Rochester in the Livonia salt shaft, there is, at the base of the Onondaga, 5 feet of coarse green and gray conglomerate containing eight species of brachiopods suggesting a mingling of the faunas of the Oriskany sandstone and Schoharie grit of the eastern part of the state.

At Honeoye Falls, 10 miles north of Livonia and 15 miles south of Rochester, the basal clastic unit is represented by 8 inches of gray calcareous sandstone without fossils.

At Oaks Corners, 41 miles southeast of Rochester, an excellent quarry exposure of the contact shows the coralliferous basal member of the Onondaga resting directly on the irregular surface of the Upper Silurian Cobleskill, with only occasional lenses of sand.

Members of the Onondaga

The Onondaga of central New York are divided into four members. In ascending order, these are the Edgecliff, Nedrov, Moorehouse, and Seneca members. The outcrop belt of the Onondaga is divided into three areas: a western area, between Buffalo and Seneca Lake; a central area, between Seneca Lake and the town of Cherry Valley, about 50 miles west of Albany; and an eastern area, between Cherry Valley and the Albany region (**Figure 8**).

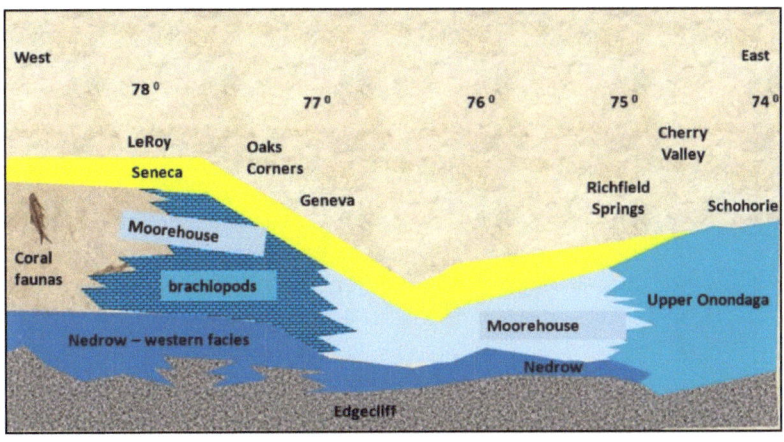

Figure 8. Generalized cross section of the Onondaga Formation from the vicinity of Buffalo to the vicinity of Albany.

Rochester lies north of the western area, with the north end of Seneca Lake about 35 miles to the southeast.

Edgecliff Member

This unit forms the basal member of the formation and is generally massive, light gray, very coarsely crystalline, and characterized by a profusion of tabulate and large rugose corals, and crinoid columnals. It has been referred to as a true biostrome.

This character applies especially to the lower half of the member. The name comes from exposures at Edgecliff Park, southwest of Syracuse. At Oaks Corners it is 10 feet thick. From Seneca Lake westward the unit thickens and light gray chert becomes more abundant. In this western area, the Edgecliff presents some striking biohermal growths. Near Buffalo, there is at least one bioherm which is about 35 feet thick and several hundred feet in diameter. Small mounds or "micro-reefs" are found commonly in this area.

From Seneca Lake eastward to Cherry Valley, the member ranges in thickness between 8 and 25 feet. The basal sand is abundant in most places ranging from a fraction of an inch to 4 feet in thickness. This zone usually contains an Onondaga fauna. Light gray chert here is found widely and irregularly spaced and its presence is more characteristic of the upper half of the member.

Eastward from Cherry Valley the Edgecliff thickens and sections from 20 to 39 feet have been reported. Light gray chert occurs irregularly distributed in the upper half of the member. The lower half exhibits especially strong biostromal characters and commonly grades into the underlying Schoharie grit. Bioherms are developed in the eastern area.

Nedrow Member

West of Cherry Valley, this member consists of thin-bedded medium gray very fine-grained shaly limestone. The type locality is a quarry 1 mile south of Nedrow. West of Seneca Lake, the member thickens. At Oaks Corners, the Nedrow is 16 feet thick, half the bulk of the member at all localities. No macrofauna is evident in the chert.

East of Seneca Lake and west of Cherry Valley the member ranges in thickness from 10 to 14 feet. The amount of chert decreases rapidly eastward from Seneca Lake so that chert is not common here. Fossils become abundant, with Platyceras, a gastropod, dominating. The lower half of the member is more shaley than the upper half. The shaly nature of the unit can be recognized as far east as Cherry Valley.

East of Cherry Valley the time equivalents of the Nedrow become less distinguishable and bear the same fauna as the Moorehouse member above. Except for the relative abundance of chert, the entire Onondaga formation in the most eastern parts is lithologically similar to the biostrome of the Edgecliff member as it occurs to the west.

Moorehouse Member

The unit is generally medium gray, and very fine grained and contains some thin shaly partings. Beds are about 2 inches to 5 feet thick. The name comes from exposures at the Onondega Prison quarry southwest of Moorehouse Flats.

West of Seneca Lake dark gray chert occurs throughout the member which ranges in thickness from 50 to 65 feet. Corals are abundant to the west. East of Seneca Lake the limestone is somewhat lighter in color and thickness decreases, ranging from 20 to 25 feet. Chert becomes more restricted to the upper half of the member, forming beds 1 to 5 inches thick. Faunal distinctions become more pronounced, with brachiopods dominating. At Oaks Corners, 36 feet of the Moorehouse are exposed.

The Moorehouse member is recognizable as far east as Babcock Hill, 12 miles south of Utica. East of this locality it cannot be differentiated from the underlying Nedrow member.

Seneca Member

Ecologically, the Seneca member marks a decline in the generally favorable conditions of the Onondaga sea. The lowest zones are most fossiliferous, each succeeding zone contains fewer fossils. Scattered nodules of dark chert occur throughout the member, and limestones are somewhat darker than those of the underlying member. Conditions improve toward the west, with facies becoming similar to those of the underlying Moorehouse member, although west of Canandaigua Lake little is known about the stratigraphy of the Seneca member.

Between the Seneca and Moorehouse members there is a remarkably uniform bed of clay more or less 6 inches thick. It is ochre colored on the fresh surface and dull gray on the weathered surface. This is the Tioga bentonite. It has been recognized in the subsurface throughout Pennsylvania and adjacent areas of West Virginia, Ohio, and southwestern New York, and is thought to be a good time marker. The bentonite has been found in quarries near Buffalo. Faunal studies have not been made to determine the character of the Seneca in this area.

West of Seneca Lake, the contact between the Seneca and the overlying Marcellus shale, where observed, is abrupt. East of Seneca Lake the contact is usually gradational. The most easterly known exposure of the Seneca occurs at Cherry Valley. Beyond this locality to the east it has not been recognized.

The upper contact, in general, the contact between the Onondaga limestone and the Marcellus shale (Hamilton group), especially in the east, is considered gradational. Certain species persist from the Onondaga into the overlying shale and others are thought to be definitely related.

Most authors subscribe to the theory of contemporaneous overlap. This produced a westward regression of Onondaga conditions so that the Seneca member thins to the east and younger beds of the Seneca are restricted to the west. There is a corresponding thinning of the lower Marcellus shale beds toward the west. It has been stated that the upper 50 feet of the Onondaga limestone in the west may be the time equivalent of the Union Spring shale (lower Hamilton) which appears as far west as Cayuga Lake.

The Tioga bentonite has not been located in the eastern area. Further surface work in eastern New York might show the bentonite in the Lower Marcellus.

Chert in the Onondaga Formation

The Onondaga of central New York contains notable quantities of chert at many exposures. At its greatest concentration, chert composes up to about one-half of the volume of the Nedrow member west of Seneca Lake.

Throughout central New York there is an eastward stratigraphic rise of dark chert accompanied by thickening of the non-cherty interval to the east. The facies relationship of the Edgecliff (light chert) and Nedrow (dark chert) west of Seneca Lake indicates that the type of chert is intimately connected with the limestone lithology. Dark chert is attributed to an excess of carbonaceous matter. This relationship between limestone and chert lithologies indicates that the causative agent in the chert formation existed during rather than after deposition of the enclosing limestone.

Chert occurs in all members exposed at Oaks Corners. It is not abundant in the Edgecliff where it appears in light gray nodules in the upper half of the member. Chert composes upward to one-half of the Nedrow member here.

The Nedrow and Moorehouse members are argillaceous; the Moorehouse presents a brachiopod facies wherein fossils can be found enclosed in chert, and in a condition of at least partial silicification. Chert of the Nedrow and Moorehouse members at Oaks Corners is characteristically dark.

The stratigraphy of the Onondaga 20 miles southwest of Rochester at LeRoy (where large quarries in the Onondaga are operated by the General Crushed Stone Company) is similar to that at Oaks Corners except that chert at LeRoy does not occur in the Edgecliff; the Nedrow may contain a little more chert; the Moorehouse is less argillaceous and contains corals. The megascopic appearance of the chert at LeRoy and Oaks Corners can be described in terms of size, external shape and spatial distribution, color and diaphaniety, and internal structure.

Most nodular masses of chert rarely exceed 3 inches in diameter. Usually they are highly irregular in outline where they are this large. However, a few nodules up to 4 inches in diameter can be noted. Nodules are usually less than 2 inches in diameter, grading down in size to small blebs. Beds of chert range usually between 1/2 and 8 inches in thickness, and between several inches and 2 or 3 feet in length. Most beds of chert are between 2 and 4 inches in thickness, and 6 inches and 1-1/2 feet in length.

External shape of the chert can be classified as follows: nodular (traced as a distorted ellipse with no re-entrants or protuberances), compound nodular (formed by coalescence of nodules), bedded (masses with large dimensions parallel to bedding, in relation to thickness), irregular (masses which present many protuberances), and brecciated (parted due to slumping before consolidation). All of these forms are arranged very commonly in zones of varying thickness parallel to bedding. Nodules may occur isolated in massive limestone beds with no apparent concentration parallel to bedding.

Chert at both LeRoy and Oaks Corners for the most part appears medium dark gray to dark gray. A very small amount of chert is light gray. Brown tints in fresh chert can be noted in chertiferous limestones which are argillaceous. In general, the darkest and lightest chert is opaque. Chert which ranges from medium light gray to medium dark gray is, in general, more or less translucent.

Among internal structures in chert, perhaps the type observed most commonly is the chert conglomerate. These occurrences present well-rounded chert pebbles enclosed in a chert matrix. Fossils are not abundant in the chert. Where fossils are particularly numerous in the limestone, associated chert may envelop some of them.

Other structures which occur in the chert are geodes up to 1/2 inches in diameter, which are usually filled or lined with either secondary quartz or calcite; irregular pits usually less than 1/8 inch in diameter, which result from the weathering of limestone inclusions in the chert; fractures which originated, for the most part, at the time the chert was deposited or during lithification of the chert. Many fractures are filled with secondary quartz or calcite.

In reference to chert at Oaks Corners and LeRoy, among evidences in support of the hypothesis of primary origin are: the concentration of most of the chert along planes parallel to limestone bedding; the presence of elliptical chert forms, which are more readily explained by a colloidal precipitation theory; the occurrence of chert conglomerate; the occurrence of cracks in some chert which are filled with limestone similar to enclosing limestone; the presence of well-preserved fossils only partially silicified in some chert masses; the absence of evidence of growth in chert since the period of deposition; the disposition of limestone laminae above and below some chert masses.

Among evidences in support of the hypothesis of replacement are the irregular shape of some chert masses; the presence of irregular inclusions of limestone in some chert masses; the association of silicified fossils and chert in some limestones; the preservation of limestone structures in some chert masses; the occurrence of silicified oolites in some limestone. The chert under discussion here can best be explained as having formed penecontemporaneously with limestone deposition.

Fossils

The fauna of the Onondaga in western New York is large and varied. The dominant group of animals were the corals which, in parts of the basal Edgecliff member, form biostromes some 20 feet in thickness. The Nedrow member overlying the Edgecliff is the least fossiliferous, whereas the Moorehouse above it has a great abundance of corals in the Rochester area. To the east, however, these corals give way to a dominant brachiopod fauna near Canandaigua Lake. Some of the more common fossils found in the Onondaga are the corals Heterophrentis prolifica, Bethanyphyllum robustus, Cystiphylloides americanum, Synaptophyllum simcoense, and Cylindrophyllum elongatnm; brachiopods Atrypa reticularisli, Chonetes mucronatus, Q. lineatus, Levenia lenticularis, and Leptaena rhomboidalis. The gastropod Platyostoma lineata is also quite common.

Many of the fossils have been silicified and, therefore, readily weather free from the enclosing carbonate matrix, or can be recovered by etching blocks in hydrochloric acid.

Facies and Faunas

Following the widespread covering of lime-depositing waters which characterized New York through most of Onondaga time, there was a basic change in the pattern of sedimentary conditions. This was primarily the result of disturbance and uplift in the east, which elevated the land area bordering the sedimentary basin, thus providing a fresh supply of clastic materials. The uplift culminated in the Acadian Disturbance which had its most intense effects in late Devonian time, but its beginnings were reflected in the sedimentary record as early as the time of deposition of the uppermost Onondaga in western New York. As noted above the upper Onondaga is thought to grade eastward into clastics of the lower Hamilton group (**Figure 9**).

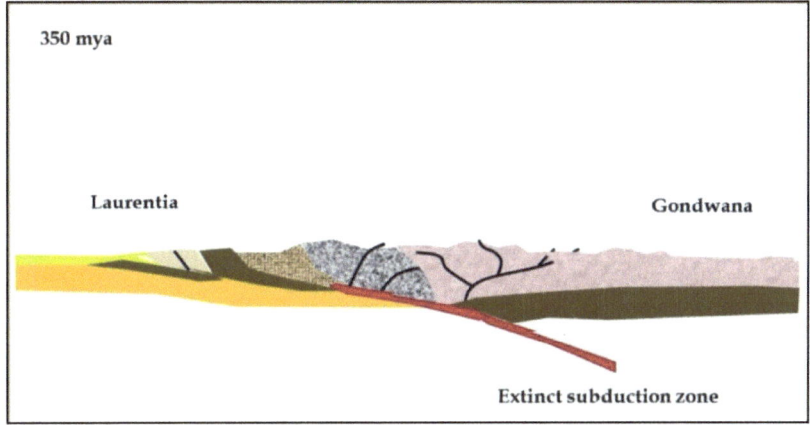

Figure 9. Cross section showing the collision occurring during the Arcadian Uplift along the continental eastern margin and Gondwana came in contact with Laurentia. The Iaepetus Ocean was completely closed at this point.

The Middle and Upper Devonian of New York are dominated by a complex series of facies. These interfingered intricately in detail, but in gross aspect they have about the relationships outlined below and indicated on the diagram. From east to west across the area at any one time the succession of facies was:
1. Continental red beds
2. Nearshore marine coarse clastics
3. Non-calcareous shales and fine sandstones
4. Calcareous shales and impure limestones
5. Black shales
6. Pure limestone (only to the west of New York State)

As time progressed, elevation of the land source in the east continued and the sedimentary facies migrated farther and farther westward as more and more clastics built out the great Catskill Delta from the east. Eventually (in the Late Devonian) the continental red bed facies reached to western New York and northwest Pennsylvania. Because of this westward migration of facies with time, a vertical section through the Middle-Upper Devonian sequence tends to show (from the base upward) the same succession of major facies as were disposed with increasing distance from the shoreline. Except for the easternmost red bed facies, marine conditions prevailed and the environment was favorable to abundant animal and plant life in many areas.

Occupying a somewhat anomalous position in the midst of this great mass of sediments is the Tully formation which includes some conspicuously pure and massive limestone beds. The sedimentary and paleogeographic setting associated with Tully deposition are not wholly understood in detail. The area of Tully outcrop lies entirely to the east of the Genesee Valleys.

The highly fossiliferous strata of the Middle and Upper Devonian have long attracted paleontologists. As might be expected, detailed study has shown that the variety of lithologic types in the sequence -- reflecting contrasting environments as they do -- are accompanied by a variety of faunal assemblages. Also, since the vast Catskill Delta required many millions of years of building, even the same lithofacies may contain distinctive faunas in portions which are of considerably different age. The approximate relationships of the lithofacies and named faunas suggests the derivation of later faunas from earlier ones. The names of faunas have been placed in quotes to distinguish them from rock-unit names.

The "Hamilton" (or "Skaneateles") fauna is the most prolific of the Middle Devonian. It shows some east-west modification depending upon the lime-clastic ratio. It is essentially a brachiopod fauna (also referred to as the Tropidoleptus fauna) with corals and crinoids abundant (westward) where the lime content is appreciable, and pelecypods added (eastward) where lime is lower and more clastics are present. The Upper Devonian "Ithaca" fauna is a recurrent "Hamilton" fauna and the "Chemung" fauna, derived from the "Hamilton", represents a near shore facies in coarse sediments and is rich in glass sponges, pelecypods and brachiopods. The "Naples" fauna includes pelecypods as the most striking element in a large and diversified assemblage with distinctive Upper Devonian forms (e.g. goniatites).

The fauna of the Tully limestone is made up essentially of the lime-loving Hamilton species (but without such abundant corals) with the addition of several exotic forms whose close affinities are with European species of about the same age.

The Middle Devonian "Marcellus" fauna of the black shale facies is decidedly limited in number of species and is typified by the brachiopod Leiorhynchus. Of course, field relationships are never so simple as can be shown on a diagram. This is especially so in the case of the Middle-Upper Devonian facies because of the vast scale and complexity of their interrelationships. Thus, individual "fingers" of a facies may be of such magnitude as to be accorded formation rank, and the vertical sequence of facies in one exposure or series of exposures may differ greatly from the diagrammed one. However, a concept of the gross arrangement of major facies may assist in understanding the stratigraphic section in the Genesee Valley and its relation to the larger regional picture (**Figure 10**).

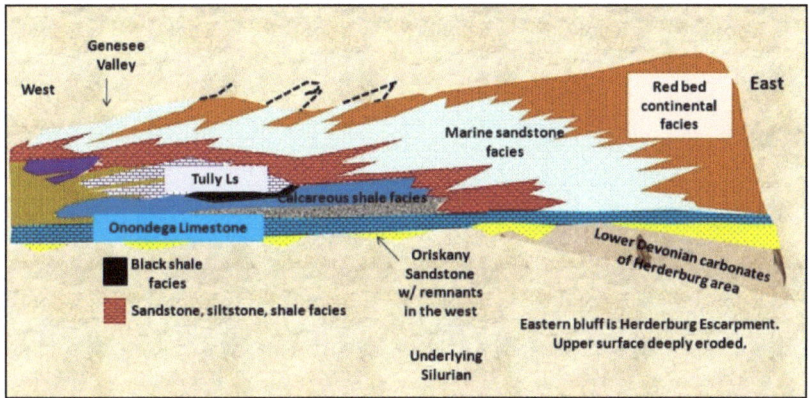

Figure 10. Highly schematized cross section with extreme vertical exaggeration suggesting major pattern of distribution of Devonian roek types across New York State.

Just as subdivisions of the lithologic facies become meaningful upon detailed study, so do subdivisions of the faunal assemblages have value. Some of the faunal groupings are discussed in the following section on the Hamilton group.

Introduction

The Hamilton group consists of a huge wedge of Middle Devonian sediments, which are dominantly subaerial and near shore clastics in eastern New York, changing over to mixed clastics and limestones in central and western New York, and finally grading into marine limestones and shales in Michigan. The rocks thin appreciably from 4,500 feet to 5,000 feet in eastern New York, to about 285 feet at Lake Erie.

In the Genesee Valley area, the Hamilton is 475 to 510 feet thick. In New York the Hamilton group represents a fascinating example of intertonguing facies with unusual and rapid changes in lithology both along the strike and vertically. Coupled with the complex lithologic pattern is the bewildering number of fossils found in these rocks. Regrettably, most of the fossils have an extended stratigraphic range and, therefore, are of little value for detailed correlation.

On the other hand, certain particular faunal assemblages have been used for correlation with moderate success. The more arenaceous eastern facies contain an unusually large pelecypod fauna, whereas the argillaceous and calcareous western facies have a dominantly brachiopod fauna with an occasional coral horizon.

The Tully formation does not extend into the Genesee Valley area. Its most western exposure is approximately one mile east of the east shore of Canandaigua Lake. At that locality, the Tully is a dense, dark gray limestone about 2 feet thick. This formation thickens to the east to about 30 feet in the Chenango Valley and the limestones grade into arenaceous shales and calcareous sandstones.

The term Hamilton was proposed for a series of shales near Hamilton, New York, that lie between the Skaneateles below and the Moscow above. This sequence is now recognized as the Ludlowville formation. Later, the term Hamilton was used to include all the rocks from the top of the Marcellus formation to the base of the Tully limestone. The Hamilton was divided into the Skaneateles, Ludlowville, and Moscow. This usage was followed until 1930 when the Hamilton was included with the Marcellus in the Hamilton. Currently, the group consists of four formations; in ascending order they are the Marcellus, Skaneateles, Ludlowville, and Moscow.

The base of the Hamilton rests on the Onondaga limestone. The nature and significance of this contact was discussed in the preceding section on the Onondaga. From Canandaigua Lake eastward to the Schoharie Valley, the Hamilton is disconformably overlain by the Tully formation, whereas westward from Canandaigua Lake to Lake Erie, it is disconformably overlain by the Geneseo formation. Evidence of this break is the successive westward disappearance of the three top faunal zones in the Hamilton and the sharp lithologic break between the Moscow and Tully formations. In addition, the magnitude of the break increases to the west as noted by the absence of all of the Moscow and part of the Ludlowville formations in Ohio and Ontario.

Descriptions of Formations and Members

All of the formations of the Hamilton group outcrop in the Genesee Valley area. The Marcellus formation consists of the Oatka Creek member, which rests directly on the Onondaga limestone, and the Stafford limestone member. Overlying the Marcellus is the Skaneateles formation composed of only one member: the Levanna black shale member. The Ludlowville formation follows the Skaneateles and is divided into the following members: Centerfield limestone, Ledyard-Wanakah shale, Tichenor limestone and shale and Deep Run shale. The youngest Hamilton formation is the Moscow which is made up of the Menteth limestone member: Kashong, and Windom shale members and the Leicester marcasite member.

The Hamilton beds consist of black shales, dark grey to blue grey calcareous and argillaceous shales and interbedded thin limestone layers and lenses. The older formations are characterized by black and dark grey shales with few limestone layers which grade imperceptibly into light grey and blue grey shales with numerous limestone layers in the upper formations. The upper part of the Moscow formation consists of dark grey to black shales and unique marcasite lenses. It is fortunate that the otherwise monotonous expanse of nearly homogeneous shales are interrupted by a few remarkably persistent limestone beds which serve as valuable key horizons across central and western New York.

Marcellus Formation

In the Genesee Valley, the Marcellus formation is represented by approximately 10 feet of black, bituminous, fossiliferous shale followed by 20 feet of dark grey fissile shale with concretionary layers and abundant pyrite. These 30 feet are represented by the Oatka Creek member. Fossils are most abundant in the upper few feet.

From its type section at LeRoy, New York, the Oatka Creek member thickens to about 50 feet at Lake Erie and to 60 feet at Cayuga Lake. The member loses its identity between Marcellus, New York, and the Onondaga Valley where it intertongues with and grades into the Cardiff shales and siltstones and the Chittenango black shale. Still farther east in the Chenango and Unadilla Valleys the Cardiff unit is divided into the Bridgewater, Solsville and Peckport members. Underlying the Chittenango shale are the Union Springs and Cherry Valley members both pinch out before reaching the Genesee Valley.

The top member of the Marcellus is the Stafford limestone, a massive dark grey limestone about 2 feet thick with shaly partings in the middle. Traced westward, the member thickens to 15 feet at Lake Erie, but eastward it thins to approximately 6 inches in the Canandaigua quadrangle. The Stafford is believed to correlate with the Mottville sandstone and limestone member to the east, but insufficient exposures prevent contInuous lateral tracing of one unit into the other. The Mottville and Stafford members were formally placed at the base of the Skaneateles formation, but in 1942 they were transferred to the Marcellus formation because of the presence of Paraspirifer in the Mottville and its abundance below the Mottville in eastern New York. In general, the Stafford limestone contains fewer species of fossils than younger Hamilton limestones.

Skaneateles Formation

This formation consists of 190 to 214 feet of undifferentiated black and grey shales, intercalated in the upper part by a few limestone and concretionary layers. The basal part consists of black shale, very similar to the Oatka Creek shale below. The black shale sequence is terminated by a grey limestone 105 feet thick. The succeeding 140 to 170 feet are composed of dark grey to medium grey shale overlain by 15 to 25 feet of dark grey to black interbedded shales with a few dense limestone bands and concretions at the top. East of Cayuga Lake, the Skaneateles formation is divided into three members, but west of Cayuga the entire formation is represented by the Levanna member. The formation thins to the west to about 43 feet at Lake Erie. Eastward it thickens to 225 feet at Canandaigua Lake and is over 400 feet thick in the Chenango and Unadilla Valleys.

The members recognized east of Cayuga Lake are the Delphi Station, Pompey and Butternut; in the Chenango Valley, the Chenango member wedges in between the top of the Butternut member and the base of the Ludlowville formation.

In the Genesee Valley the Ludlowville formation is 115 to 130 feet thick. The base is marked by several very fossiliferous limestone beds. Above the limestones, the shale is very dark, but gives way to lighter shales and a few thin limestone beds in the upper part. Four members are recognized in this area. They are, from the oldest, the Centerfield, Ledyard- Wanakah, Tichenor and Deep Rung Centerfield limestone members (**Figure 11**).

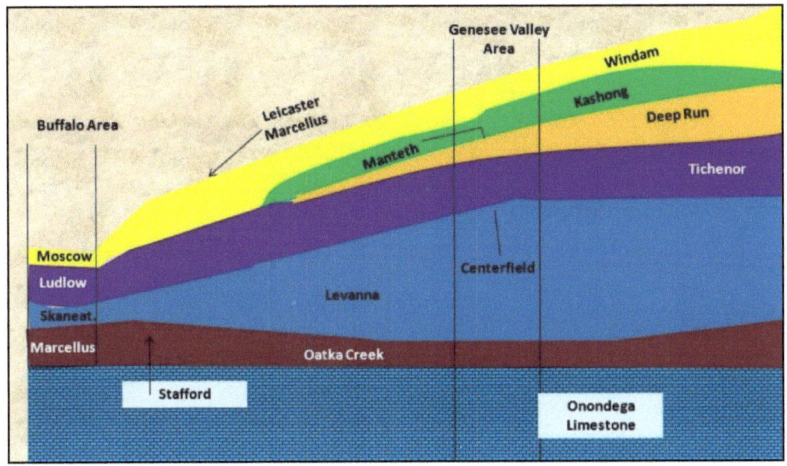

Figure 11. Cross section of the Hamilton group from Buffalo through the Genesee Valley eastward.

This unit consists of 8 to 11 feet of five to seven dense crystalline extremely fossiliferous limestone beds separated by several inches of blue grey shale. The Centerfield is famous for its huge and varied fossil assemblage, of particular importance are its vast coral coloniese. This member thins westward to 405 feet at Blossom, New York, and is believed to correlate with the Pteropod zone of Grabau along Lake Erie.

At Skaneateles Lake, it is 50 feet thick. The Centerfield becomes homogeneously arenaceous and cross bedded east of the Finger Lakes and loses its identity in the Chenango Valley. The Centerfield is tentatively correlated with the Stone Mill limestone to the east.

Ledyard-Wanakah shale member

This interval is divided into two members separated near the middle by the Strophalosiali (Productella) bed. In the Batavia quadrangle the "Strophalosia" bed does not maintain a constant stratigraphic position with respect to the overlying Tichenor limestone, lying there only 1.5 feet below the Tichenor. Aside from the Strophalosiali bed there is little on which to separate the Ledyard and Wanakah.

The member is 105 to 112 feet thick in this region. The lower 90 feet is composed of dark grey shales interfingered with grey to black fissile shales, thin fine-grained limestones and concretions. This sequence is overlain in the Batavia quadrangle by a light grey argillaceous limestone 0.5 to 100 foot thick which contains abundant Productella truncata (formerly Strophalosia truncata). Above the "Strophalosia bed" the shales are light to medium grey and intercalated with numerous thin limestone layers and lenses. Fossils are only moderately common in the shales below the "Strophalosia" bed and are confined to a relatively few species.

In contrast, the shales above that bed are generously fossiliferous and contain a greater variety of species. This is the first occurrence of what is termed "the typical Hamilton fauna".

Tichenor limestone member

The Tichenor is made up of 8 to 11 feet of alternating semi-crystalline blue grey limestones and calcareous medium grey shales. Both are highly fossiliferous and corals are particularly abundant.

Deep Run shale member

The Deep Run member is represented by about 9 feet of brittle blue to blue-grey shale. These rocks carry a fauna quite distinct from the underlying Tichenor member by having pelecypods and brachiopods in abundance but by almost lacking corals and Bryozoa.

Post-Centerfield correlations

All of the Ludlowville members recognized in the Genesee Valley thin to the west. At Lake Erie, the Ledyard-Wanakah is 7.5 feet thick and the Tichenor less than 2 feet. The Deep Run is not known west of the Batavia quadrangle. To the east, the members thicken. The Ledyard-Wanakah is 130 feet at Cayuga Lake, the Tichenor is 11 feet at Canandaigua Lake but thins to I foot at Seneca Lake. The Deep Run is 55 feet at Canandaigua Lake but thins to 49 feet at Seneca Lake.

The lower part of the Ledyard-Wanakah sequence grades into the Otisco Member, and the upper part of the Ledyard-Wanakah, the Tichenor and Deep Run members grade into the King Ferry arenaceous shale in the Cayuga Lake region. In the Chenango and Unadilla Valleys, the Ludlowville is undifferentiated.

Moscow Formation

The Moscow formation in the Genesee Valley consists of 9.5 to 10.5 feet of blue-grey to dark grey shales intercalated with numerous limestone bands and concretionary layers. Four members are recognized within the formation. These are, in ascending order, the Menteth limestone, Kashong and Windom shales, and the Leicester marcasite (Figure 12).

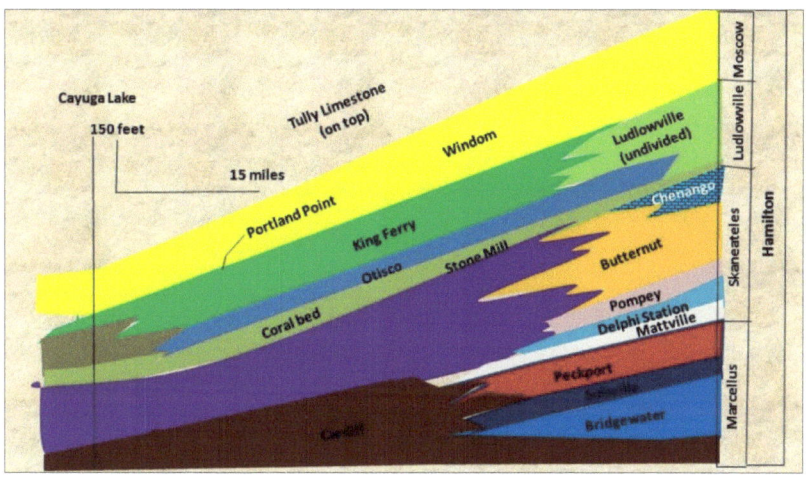

Figure 12. Cross section of the Hamilton Group from Cayuga Lake eastward.

Menteth limestone member

This member is a 0.5 to 1.0 foot dense, coarsely crystalline, crinoidal limestone. It is abundantly fossiliferous. At some exposures the member is corniferous, and some of the fossils are silicified. In addition, the fossils and matrix frequently have a pinkish sheen. This member thins to the west and terminates in the western part of the Batavia quadrangle. To the east, it is correlated with the basal beds of the Portland Point member.

Kashong shale member

The Kashong member is composed of 45 to 55 feet of blue and blue-grey shale, very similar to the Deep Run shale, with concretions and limestone layers in the upper part. The lower 10 feet consists of blue-grey shale with numerous crinoid fragments and Bryozoa. In this interval the fossils decrease in abundance toward the top. The succeeding 26 to 36 feet of shale is blue or blue-grey and contains a varied fauna, but specimens are not abundant with the exception of Tropidoleptus carinatus and Phacops. The remaining 9 to 10 feet is distinguished by three limestone-concretionary beds, each separated by approximately 4 feet of blue shale which commonly contains small pyrite nodules and oddly shaped concretions.

The top of the upper limestone layer is regarded as the Kashong-Windom boundary in this area, rather than the proposed Ambocoelia-Chonetes zone. The reason for this is explained in the discussion of the Windom shale. The Kashong member is a large lenticular mass that thins east and west of the Genesee Valley, and extends from slightly east of Cayuga Lake to the eastern part of the Depew quadrangle.

Windom shale member

The Windom member is made up of 45 to 48 feet of shale and thin interbedded limestones. The lower 12 feet of shale is fine grained and medium grey. The rocks in the upper 5 feet of this interval contain enormous numbers of the small brachiopod Ambocoelia umbonata. The following 8 to 12 feet of shale is darker, and carries a more varied fauna. Above this is a short interval of dark grey shale that contains Leiorhynchus, and Ambocoelia praeumbona. The succeeding 20 feet consists of interbedded thin coquinoid limestones and very fossiliferous medium grey shales. Numerous concretions are imbedded in the shales.

The upper 4 to 5 feet of shale is dark grey to black and carries few fossils except Leiorhynchus laura and Lingula sp. The Ambocoelia umbonata-Chonetes mucronatus zone was considered to be the base of the Windom. Recent studies indicate that west of Canandaigua Lake there is not one, but several layers in the Windom that contain abundant Ambocoelia umbonata and Chonetes mucronatus. In addition, these very fossiliferous layers do not maintain a constant stratigraphic position with respect to the top of the formation.

It was suggested that the first limestone below the lowest occurrence of Ambocoelia-Chonetes be used as the division between the members in this area. This limestone is recognized in the Batavia and Attica quadrangles, and is tentatively correlated with the limestone layer below the Ambocoelia-Chonetes zone at the type section of the Moscow. The Windom member is 15 feet thick along Eighteen Mile Creek, 135 feet at Cayuga Lake and 265 feet in the Unadilla Valley.

Leicester marcasite member

From Canandaigua Lake to the western edge of the Depew quadrangle the Windom is usually separated from the Geneseo black shale by thin lenticular masses which consist chiefly of marcasite and pyrite with minor amounts of calcium carbonate. They average 5 inches at their thickest part, and from 20 to 30 feet in length. These unusual rocks contain a dwarfed fauna which is difficult to free from the dense, hard matrix. In earlier reports this unique horizon was referred to as the "Tully" pyrite because of its stratigraphic position. As the name implied, it was regarded as the western equivalent of the Tully limestone.

The "Tully" pyrite was correlated with the Spirifer tullius-Vitulina (now Pustulina) zone, i.e., the highest faunal zone in the Hamilton in central New York. To remove any suggestion of correlation with the Tully, the name Leicester marcasite member was proposed for the lenses, and it was considered to be the youngest member of the Moscow formation. More recent work in the Canandaigua Lake area suggests that the marcasite lenses may be post-Hamilton - pre-Tully in age, and there is a possibility that they may even be post-Tully in age. However, a detailed study of the marcasite over a larger area is required before a more definite age determination can be made.

Upper Devonian

Genesee Group

The Geneseo Black Shale

The Geneseo shale is considered to represent the basal formation of the Genesee group, which is lowermost Upper Devonian in age. The lower part of the formation consists of black, finely laminated shale. The fresh rock is massive but upon exposure fissility is developed. Upon breaking open a fresh block of this shale, the observer is at once aware of a strong petroliferous odor. Overlying these beds is a series of dark grey, irregularly bedded shales, thin limestones, and concretions. To the east, at Cayuga Lake, the upper portions contain thin, cross-bedded, light-colored siltstones.
At Lake Erie the Geneseo is 2 to 8 inches thick. There are 84 feet of the formation at Fall Brook (type section, 1.75 miles south of the town of

Geneseo on the east side of the Genesee Valley). At Canandaigua Lake 115 feet of Geneseo has been measured, while at Cayuga Lake 125 feet is reported (**Figure 13**).

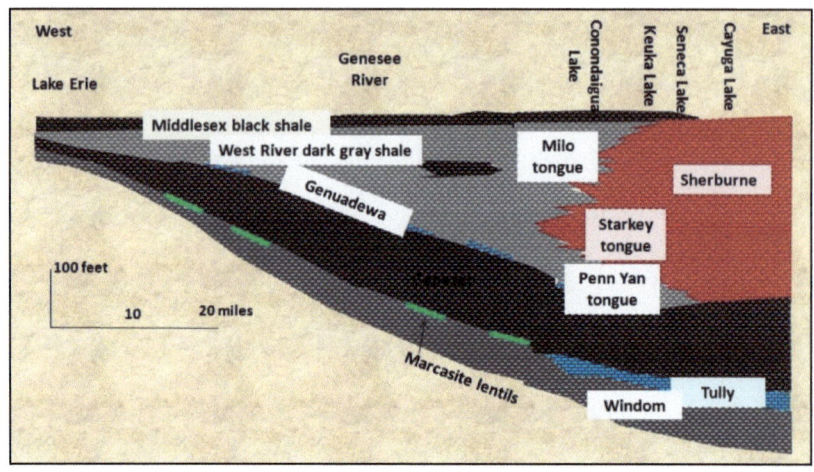

Figure 13. Generalized facies relationship of the Genesee Group from Lake Erie to Cayuga Lake.

Genundewa Limestone Lentil of the West River Shale

The Genundewa limestone type section is located at the cliff-shore exposures positioned at the foot of Bare Hill (formerly Genundewa Hill) on the east side of Canandaigua Lake. The Geneseo black shale is separated from the main part of the West River shale by a series of thin limestones and shale beds referred to as the Genundewa limestone. The base of the Genundewa is defined as the first Pteropod limestone in the Geneseo-West River sequence which contains the pelecypod Paracardium doris and/or the goniatites Manticoceras spo and Tornoceras cr. Touniangulare. The top of the Genundewa cannot be defined because the characteristic thin limestones are found at varying distances (up to 46 feet) above the base of the West River shale.

Variations in thickness are due to the intervening shale. Because of the vagueness of the upper boundary of the member, the term Genundewa was abandoned. These limestone lentils were called the Styliolina beds, considering them as a facies of the West River.

The Genundewa consists of dark to light brownish gray, lenticular, nodular and concretionary limestones which contain an abundance of the shells of Styliolina fissurella. This tiny, needle-like pteropod may be seen by a close look at most Genundewa specimens. The thin limestones are separated by irregularly bedded, dark grey shales which have a typical West River aspect. The Genundewa extends from Lake Erie where it is only a few inches thick to the vicinity of Gorham, Ontario County, or possibly to Seneca Lake.

The fauna of the Genundewa represents the first appearance of the Naples fauna which characterizes the formations above the West River shale. The abundance of the pteropod Styliolina fissurella has been mentioned. The other common forms include the ammonoid cephalopods (e.g., Manticoceras sinuosus and Tornoceras uniangulare), thin-shelled pelecypods (e.g., Paracdium doris), and plant remains.

West River Shale

The West River shale was named for the sequence of shales above the Genundewa and below the Middlesex black shale. Favorable exposures in the West River Valley of Yates County were chosen as the type section. The major portion of the West River consists of interbedded dark grey and black shales. The dark grey units are irregularly bedded and often calcareous. The black shales are fissile and have the appearance of the Geneseo black shales.

At Lake Erie, the West River as a whole (including the basal Genundewa limestone lentils) is 8.5 feet thick. At the Genesee Valley it is 60-70 feet thick and at Canandaigua Lake it is 175 feet thick. Between Canandaigua Lake and Cayuga Lake, the West River thins to 36 feet. In the Canandaigua Lake area the West River is penetrated by the Starkey tongue of the Sherburne sandstone and is divided into the Penn Yan tongue (lower) and the Milo tongue (upper). The Penn Yan tongue is the part of the West River which extends to Cayuga Lake; the Milo tongue is replaced by the Starkey prior to reaching the east side of Seneca Lake. In the Canandaigua Lake-Beneca Lake area the Starkey tongue consists of thin, light colored, cross-bedded siltstones interbedded with the dark grey shales of the West River.

The first continuous sequence of black, fissile shale above the West River defines the base of the Middlesex black shale. The West River is not abundantly fossiliferous; thin-shelled pelecypods and brachiopods are its most common fossils. The pelecypod pterochaenia fragilis is one of the most abundant of these.

Sedimentary Structures in the Genesee Group

Concretions and septaria are of particular interest in the West River shale and Geneseo black shale. They consist of dark grey to greyish black, argillaceous limestone and in many places occur in rows. They are commonly spheroidal in shape but may be quite elongated and irregular. Bedding may be continuous from the surrounding grey shale through the concretion. Most commonly the concretions contain concentric layers. Peripheral haloes of finely disseminated pyrite have been observed in the Geneseo concretions. Often the concretions contain a pyritized nucleus, a fossil animal or plant fragment, with the shape of the concretion influenced by the shape of the nucleus. Solid limestone concretions may contain a single thin crack-wedge filled with crystalline mineral material. By an increase in the number of these cracks, concretions grade into septaria.

Septaria are most commonly found in the West River shales. Contained in the wedge shaped cracks of these rounded bodies are crystalline mineral suites which occur in vugs in some of the thick limestone and dolomite formations of the Silurian (Lockport dolomite is an example). Small euhedra of pyrite, ankerite, sphalerite, galena, selenite, and, most commonly, calcite are often found lining the V-shaped fractures. The origin of this very interesting sedimentary structure is not completely understood. The concentration of a metallic mineral suite in a ball of impure lime mud and the concentration of the lime mud into spherical shapes is an unsolved problem in sedimentation which may be intimately related to the origin of limestone. The shales surrounding concretions and septaria are commonly bent around these structures. This strongly suggests a contemporaneous time of deposition for the shales and the concretions.

Lobate flow markings on the base of siltstone beds are commonly found in the West River shale. Parallel flow markings are found on siltstones in the Geneseo black shale. These two structures are very common in the Naples group.

Naples Group

Introduction

The Naples group consists of four formations, in ascending order: the Middlesex black shale, Cashaqua formation, Rhinestreet black shale, and Hatch formation. The group extends from Seneca Lake to Lake Erie and thins from east to west. Sandstones, siltstones, and gray shales predominate in the eastern part. In the west, dark gray and black shales are characteristic. The name Naples was first used for the strata above the Genesee shale.

The usage of the term was revised, limiting it to the rocks between the West River (below) and the Grimes sandstone (**Figure 14**).

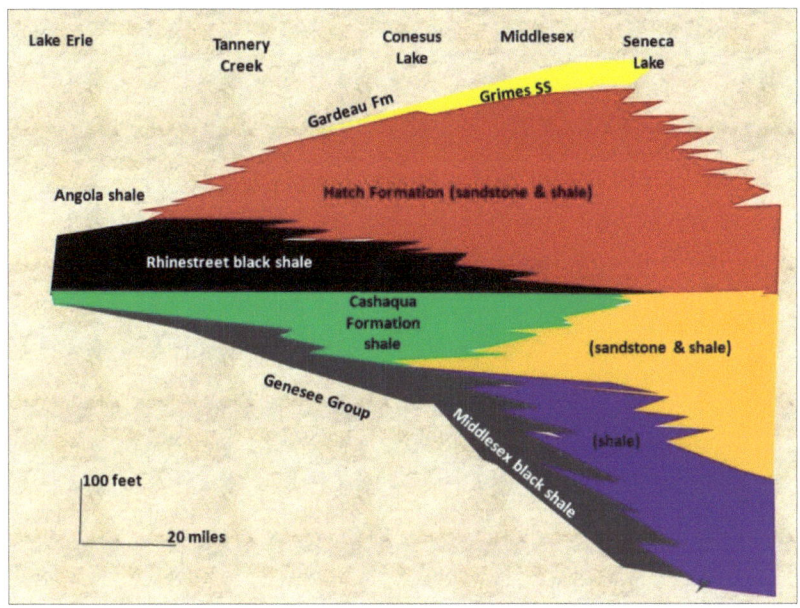

Figure 14. Facies relationships in the Naples Group.

The Naples group is approximately 500 feet thick in the Genesee Valley. Westward, it thins to 170 feet at Lake Erie. Eastward, it thickens to over 800 feet at Seneca Lake and beyond its identity is lost where it interfingers with the Ithaca and Enfield formations at Cayuga Lake.

Middlesex Black Shale

"Middlesex" was applied to the black shales at Middlesex, New York. The formation is 30 feet thick in the Genesee Valley and is composed of black shale with a few thin beds of gray shale. Fossils are very rare. The formation is exposed just above the base of the dam at Mount Morris. It may be recognized by its dark gray color, blocky jointing, and resistant character.

Cashaqua Formation

The name, Cashaqua, was given to the shales overlying the Middlesex black shale. The type locality is Keshaqua Creek (formerly Cashaqua Creek), southeast of Mount Morris. The formation thins from east to west, measuring 480 feet at Seneca Lake, 166 feet at Mount Morris, and 28 feet at Lake Erie. Bluish gray shale (weathering to olive-gray) predominates in the Genesee Valley-Lake Erie region.

The formation becomes more sandy toward the east. At Seneca Lake, it is composed of 40 percent sandstones and siltstones and 60 percent arenaceous shales.

The Cashaqua forms most of the gorge wall at the Mount Morris dam. The upper contact may be observed a short distance below the top of the cliff. Fossils are common. Thin-shelled pelecypods, cephalopods, and gastropods predominate. Brachiopods, arthropods, crinoids, and fish are rare It.

Rhinestreet Black Shale

"Rhinestreet" was given to the black shales overlying the Cashaqua formation. The type locality is just north of Naples, New York. The formation is composed of black shale interbedded with small amounts of dark gray shale and thin siltstones. It thickens from east to west and measures over 100 feet in the Genesee Valley. Fossils are rare in the black shale but a Cashaqua fauna may be found in some of the dark gray shale beds in the lower part of the formation.

Hatch Formation

The Hatch formation was named for the type locality designated as the lower part of Hatch Hill, near Naples. The formation extends from Attica, New York, to Keuka Lake. In that distance, it thickens from 150 feet (west) to over 400 feet (east). At Letchworth State Park it measures 186 feet. Thin bedded siltstones, gray and black shales comprise the formation in the Genesee Valley. Shales are more abundant to the west, whereas siltstones predominate in the east.

The Hatch is exposed north (downstream) of the lower falls in Letchworth Park. Fossils are not common in the Hatch. A modified Cashaqua fauna is present but in greatly reduced numbers.

Formations above the Naples Group

Grimes Sandstone

The Grimes sandstone was named for exposures in Grimes Gully at Naples, New York. The formation is 25 feet thick in the Genesee Valley, and is composed of bluish gray sandstones, siltstones, and arenaceous shales. It may be observed just below the lower falls at Letchworth State Park. The Grimes carries a Chemung fauna with brachiopods predominating. A well-defined faunal break marks the Hatch-Grimes boundary.

Gardeau Flags and Shales

The name was first used for exposures on the Gardeau reservation, south of Mount Morris, New York. It included the rocks from the base of the Rhinestreet to the top of the Nunda sandstone. The strata is restricted in use between the Grimes and Nunda sandstones. This definition is followed here. The formation consists of bluish gray sandstone, siltstone, and gray and black shale. The Gardeau measures 344 feet in Letchworth Park and is exposed in the river from the base of the lower falls to a point just below the crest of the upper falls. The Gardeau and the Hatch form the walls of the gorge north of the lower falls. Fossils are uncommon. Thin-shelled pelecypods, cephalopods, and gastropods occur in some shale beds.

Nunda Sandstone

The name "Nunda" was proposed for the thick beds of bluish gray sandstone and thin beds of gray, arenaceous shale that overlie the Gardeau. The greater thickness of the sandstone beds (up to 15 feet) distinguishes the Nunda from the Gardeau below. The formation is approximately 200 feet thick in the Genesee Valley and forms the cap rock of the upper falls at Letchworth Park. Fossils are rare. A few cephalopods (Manticoceras), Aulopora, Orbiculoidea, and crinoid stems have been reported. Scolithes verticalus (a worm tube?) is very abundant.

Wiscoy Flags and Shale

The Wiscoy was first named for the soft shales that overlie the Nunda sandstone. The type locality is Wiscoy Creek, Allegheny County. The formation is 170 feet thick in the Genesee Valley and contains a black shale in the lower part called the "Pipe Creek member". Gray shales and thin siltstones comprise the remainder of the formation. A thin-shelled cephalopod and pelecypod is present in the gray shales.

Chapter 4. Quaternary Glaciology

Glacial Geology

Chronological Wastage of Ice in New York State

The Wisconsin glaciation is the major and only important glacial stage represented in New York State. Aside from the terminal moraine, all morainal deposits here are recessional, thereby indicating periods of stagnation as the ice retreated northward across the state.

Terminal or Olean-Salamanca moraine

The farthest advance is evidenced by the massive terminal moraine or Olean-Salamanca moraine which at its eastern end extends from Long Island westward through New Jersey and Pennsylvania. It swings northwest from Pennsylvania into southwestern New York in the Olean-Salamanca districts, and then turns southwest again back into Pennsylvania. That portion of the terminal moraine in New York State represents an ice front re-entrant. Patches of Illinoian drift have been found just south of the moraine in the Salamanca district.

Binghamton moraine

The Binghamton moraine extends southwestward from the northern tributaries of the Susquehanna Valley and the Chenango River Valley to the Binghamton area, and then swings northwestward to just north of Franklinville, where it turns to the southwest and continues on into Pennsylvania. This moraine has been placed in the Cary substage of the Wisconsin.

Valley heads moraine

An uneven recession of the ice front from the terminal moraine to a line of stagnation just below the southern extremities of the Finger Lakes, or on a line along the heads of the present north-trending valleys, permitted an abundance of glacial drift to be deposited. Drift accumulation in the valleys exceeded that on the intervening ridges and lines of drift across the ridges are poorly developed and difficult to trace.

In the Genesee Valley, this drift is responsible for shunting the river from Its preglacial channel and consequent excavation of the upper Genesee gorge. The Valley Heads moraine constitutes the present divide between north and south drainage, except for the major flow of the Genesee River. Like the preceding Binghamton moraine, this moraine has been dated as Cary.

Hamburg-Batavia-Victor moraine

Another recession and period of stagnation caused considerable drift deposition extending northeast from Hamburg to Batavia, where the belt turns southeast and on to Victor. The moraine continues to the east at approximately the northern ends of the Finger Lakes in the central portion of the state.

An outstanding kame area to the southwest of Victor occupies about 20 square miles and attains summit elevations up to 1000 or 1100 feet. Along the line of the moraine, the inter-valley ridges show a paucity of drift and no definite lines of ice front accumulation (**Figure 15**).

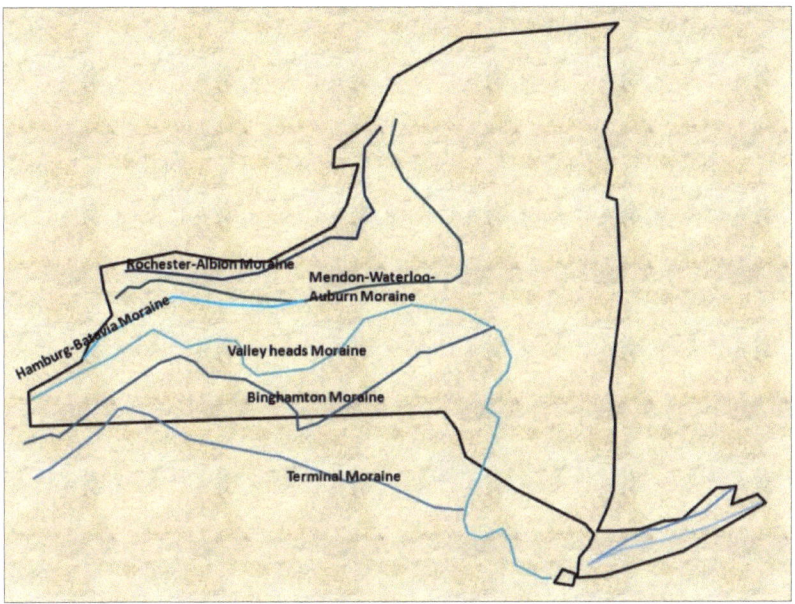

Figure 15. Moraines in Western New York and adjoining areas.

The idea was put forth that the Batavia deposits mark a point of turning or pivot of the Ontarian ice bodies.

Mendon-Waterloo-Auburn moraine

This morainal belt intersects the preceding one at Waterloo, to the west it lies sub parallel to and a few miles north of the earlier one. The Mendon kame area is about 10 miles south of Rochester. Westward from the Mendon kames, the moraine stretches to the northwest in a series of moderate knolls and short ridges interrupted only by the Genesee River. The Turk-Baker Hills, 7 miles east of Mendon, and the Junius kames between Lyons and Geneva, are included in this moraine.

To the east of Seneca Falls and Cayuga Lake, the moraine follows on line to Auburn. Significantly, the moraine is apparently contemporaneous with attenuated drumlin flutings to the north. A similar relation is described east of Cayuga Lake in the Auburn moraine. The age of the drift is Mankato.

Rochester-Albion moraine

The Rochester-Albion moraine extends west from Rochester, then curves northwestward and again westward through Albion and Medina. Along much of its course this moraine is a belt of subdued, irregular knolls and ground moraine. The line of prominent kames which constitute the Pinnacle Range in Rochester is the eastern part of the Rochester-Albion moraine.

This conspicuous ridge extends about four miles eastward from the Genesee River along the southern edge of the city. The abrupt eastern termination is imperfectly understood. There may be correlation between this belt and another one east of the Irondequoit Valley, but no direct connection exists. The Rochester-Albion moraine represents a period of recession and stagnation of the Mankato ice mass.

Oswego moraine

Near the shore of Lake Ontario a prominent ridge of water laid deposits extends from Fairhaven to Oswego, except for a break across the deep embayment of Sodus Bay. The moraine strikes essentially parallel to the shore of Lake Ontario. It was deposited in front of Mankato ice.

Glacial Lake Development in Western New York

As the ice sheet retreated from the higher elevations of its farthest southward advance to lower and lower elevations toward the north, glacial melt water and north flowing streams became impounded between the highlands to the south and the massive ice dam to the north.

The ice mass of western New York was manifested as tongues or lobes which flowed along paths of least resistance, filling and modifying valleys and leaving intervening ridges either exposed to the atmosphere or under a relatively thin sheet of ice. As the ice retreated northward, the major pre-existing north-trending valleys were uncovered, becoming the loci for the accumulation of glacial melt water and debris.

The withdrawing ice uncovered lower and lower outlet channels that allowed waters to flow east or west contiguous with the front of the ice, cutting prominent east-west channels. The east-west channels are especially well developed in the intervening ridges between and the drainage systems and lakes of west and central New York State. The ice front channels are readily seen in the field and are just as important as the moraines in defining the position of the ice front.

Many have been used for correlation purposes where morainal material is absent. Such ice border drainage was responsible for the dissipation and removal of much glacial drift.

The succession of glacial lakes in western New York is denoted by elevations of channel outlets, terraces, beach deposits, and deltas, the last being built by streams flowing in from the east or west off the inter-valley ridges. As previously stated, early glacial lake formation in the southern portions of the state was confined to the major valleys. The recession of the ice to parallels between the Valley Heads moraine and the Hamburg-Batavia-Victor moraine was accompanied by a gradual merging of the confined lakes into a large glacial lake extending many miles east and west.

The following discussion outlines the succession of major glacial lakes from the time when the ice was at the latitude of the Finger Lakes to the time of its retreat to the Ontario basin and Canada, encompassing Cary and Mankato time. The southerly, higher, and more local glacial waters were confined to the major depressions of the Genesee Valley and the Cayuga basin. Early outflow was to the south. At a later stage when the ice was at or near the northern end of the Finger Lakes, the waters collected mainly into two large lakes, which later merged to form Lake Newberry.

Lake Newberry

Lake Newberry occupied the central valleys of Seneca, Cayuga, and Keuka and discharged southward through the site of the community of Horseheads (900 feet elevation) to the Chemung and Susquehanna Rivers.

The other main lake in the Genesee River valley escaped at different times and levels via the Susquehanna-Alleghany-Ohio-Mississippi drainage. When the ice front receded to the parallel where the Hamburg-Batavia-Victor moraine now lies, at the northern end of the Finger Lakes, the two lakes merged to be called, collectively, Lake Newberry. The lake waters had a surface elevation of 1000 feet.

Lake Hall

It was mentioned under the discussion of moraines that the Batavia portion of the Hamburg-Batavia-Victor drift belt marked a point of "pivot" in the retreat of the ice front. While the ice at Batavia remained stationary, the eastern extremity of the ice front retreated and permitted Lake Newberry to drain eastward via the Mohawk Valley.

This caused lake levels to drop from 1000 to 900 feet. Glacial waters at the interval between 1000 and 900 feet were called Lake Hall.

Lake Warren I (Lake Vanuxem I)

After a time the ice front at the longitude of present Batavia receded to the Onondaga escarpment and the surface elevation in the lake dropped to just below 900 feet. This allowed the Warren waters in the Erie basin, previously trapped by the Batavia salient, to flow eastward via the Mohawk Valley. The new lake was called Lake Warren I (Vanuxem I).

Period of free drainage or deglaciation

The ice at this time apparently receded some distance to the north, a period of "free drainage" ensued and Lake Warren I disappeared.

Lake Warren II (Vanuxem II)

A subsequent advance of the ice to the south following the period of free drainage blocked off the Syracuse channels on the east resulting in the formation of a new lake, Lake Warren II (Vanuxem II). This re-advance of the ice is marked by the Mendon-Waterloo-Auburn moraine which shows the wave work of Lake Warren II at 880 feet.

The Bristol Hills, south of Mendon Ponds Park marks the Appalachian Plateau rising up from the Interior Lowland. The hills were the southern shores of Lake Warren II as evidenced by terraces and delta deposits at 880 feet.

A study of the depth of leaching in glacial gravels near Syracuse revealed anomalous data which have been attributed to an advance and over-riding by the Mankato ice sheet over a pre-existing drift. This information tends to corroborate the ice advance to establish Lake Warren II.

Lake Dana

During the recession of the ice front parallel to Rochester, water levels dropped as lower channels were uncovered. When the ice front stagnated along the Rochester parallel, glacial Lake Dana (Lundy) came into existence. The lake was dammed on the south by the Bristol Hills with the Mendon Ponds kames projecting above water level as numerous local islands. Dana lake levels have been recorded at 700-725 feet.

The southwestern flanks of the Turk-Baker Hills, 7 miles east of Mendon Ponds Park, show delta plains at an elevation of 720 feet that were built in Lake Dana. Debris-laden melt waters flowing from the ice front into Lake Dana were responsible for building the Pinnacle Range of kame deposits in present Rochester.

Great quantities of fine sediment were deposited farther out into the lake, developing the present large clay plain south of the Pinnacle Range. Excavations for buildings for the University of Rochester River Campus and the Strong Memorial Medical Center revealed great thicknesses of silts and clays. The Marcellus-Cedarville channel in the Otisco Valley far to the east is the only prominent channel that can be correlated with the Dana lake level.

Lake Scottsville

Lake Scottsville was the successor to Lake Dana and contemporaneous with the initial stages of Lake Dawson. This small, shallow, and local lake was located south of the Pinnacle moraine and west of a drift ridge (East Henrietta ridge) extending to the south along South Avenue and the East Henrietta Road.

The outlet was northward through the moraine where the Genesee River now flows. It is assumed that the lake waters cut through the lowest point of the Pinnacle Range. A contour map shows that the breach across the moraine could not have been much above 540 feet, otherwise the cut would have been made towards the west across Brooks Avenue.

Lake Scottsville extended south up the Genesee Valley past Scottsville toward Avon at an elevation of 540 feet. This lake furnished a basin for the accumulation of sediments brought down by the Genesee River. The expansive valley plains from Rochester to Avon are lake sediments topped by silts left by river floods.

The Lake Scottsville waters discharged northward through the Pinnacle Range into the beginning stages of Lake Dawson. Drops in the level of Dawson caused Scottsville waters to cut a rather prominent channel through which the Genesee River now flows.

Lake Dawson

Lake Dawson followed Lake Dana in the sequence of temporary glacial lakes and existed for a relatively short period of time. The lake occupied the west end of the Ontario basin and flooded the Irondequoit Valley. A portion of the ice front along the east side of Irondequoit Bay and along the parallel of Penfield was responsible for the western restriction of the lake (**Figure 16**).

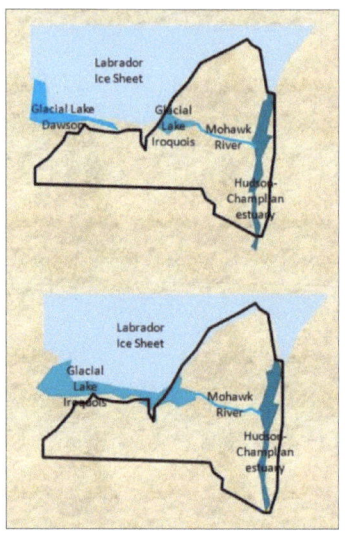

Figure 16. Two stages in glacial retreat showing Lakes Dawson and Iroquois.

Lake Dawson drained to the east via a capacious channel commencing at Fairport and passing through Palmyra, Newark, Lyons, and Clyds. The lake ultimately discharged into the incipient stage of Lake Iroquois then inundating the districts from Rome to Syracuse and the eastern Ontario basin. The extinction of Lake Dawson took place when the ice, lying along the parallel of Penfield, receded and allowed the waters of Lake Dawson and the early, eastern Lake Iroquois to merge, forming greater Lake Iroquois.

The pro-glacial lakes of Dawson and Iroquoit completely filled and overtopped the ancient Irondequoit Valley. This resulted in a copious and almost complete filling of the valley by sediments which in turn lie over a considerable accumulation of till.

The elevation of Lake Dawson has been placed at 480 feet. This figure was reached by considering a 460 foot elevation at the Fairport channel, a 15 foot allowance for depth of water in the channel, and a 5 foot difference of land uplift between Fairport and Rochester. During Lake Dawson time, Niagara Falls and Lake Erie came into existence.

Lake Iroquois

The long duration of Lake Iroquois resulted in a well-defined shoreline and conspicuous wave-cut cliffs. The lake level was determined by the elevation of a pass at Rome leading into the Mohawk Valley. In the Rochester district, the Iroquois shoreline is found along the 440 foot contour. Ridge Road runs along a prominent Iroquois beach just to the north of the city proper. The lake inundated the Irondequoit Valley as far south as Pittsford. Processes of sedimentation in the Irondequoit Valley were extensive during Lake Dawson and Lake Iroquois times, thereby filling the depression to within 30 or 35 feet of the water surface. More recently, Irondequoit Creek and distributaries have contributed to fill in the deeper parts of the valley, while, at the same time partially removing some of the lake deposits from higher levels.

When removal of the ice opened the St. Lawrence River Valley, Lake Iroquois was slowly drained to sea level. The ephemeral sequence of falling Iroquois waters left behind a series of terraces which can be readily observed in the Irondequoit Valley. These have been modified by more recent erosive effects of the Irondequoit Creek and its tributaries.

Gilbert Gulf

Draining of Lake Iroquois ultimately reduced its surface elevation to sea level. The name applied to the ocean level water in the basin of Lake Ontario is Gilbert Gulf. Since Gilbert Gulf time, isostatic uplift has caused the rise of the surface of Lake Ontario to its present 246 feet above sea level.

Mendon Ponds County Park

The Mendon kame district is approximately 10 miles south of Rochester. It is the third highest point in Monroe County, with a maximum altitude of 840 feet. It is bounded on the west by Clover Road and on the east by Pittsford Road. The park covers 4 square miles. Here is an excellent example of a kame-kettle topography, similar to the western extremity of the Pinnacle Hills at Mount Hope Cemetery. The map of Mendon quadrangle shows the district to have a three-fold east-to-west division.

The central division is an area of impounded drainage exemplified by four major lakes or kettles. The eastern and western divisions each contain a well-developed esker flanked by prominent kames. The esker along the western tract is approximately 2 miles long and reaches a height of 100 feet at several places. It abruptly broadens into an esker fan at its southern termination west of Deep Pond.

The esker extending north-south along the eastern division is one of the best esker forms in New York State. It possesses a typical hummocky profile and assumes a serpentine course. The esker is bordered on both sides by kames, and extends in a northeast to southwest direction for a distance of 2-1/2 miles. A succession of numerous kettles occurs along its base, which may contain water forming small ponds or swamps. The northern end of the esker is one of the highest points in Monroe County. The esker becomes conspicuously subdued at its southwestern terminus.

The Turk-Baker Hills, 7 miles east from Mendon Ponds Park are also kame deposits and attain an altitude of 930 feet. The Turk-Baker Hills display wave work done by Lake Warren II at just under 900 feet. On the southwest flanks of the hills are delta plains built in Lake Dana at 700 feet. The higher kames at Mendon display some inconspicuous leveled or smooth tracts on the east and west side of the kame areas where Dana waves had greater force.

The Junius kame moraine 20 miles southeast of the Turk-Baker Hill kames is correlated with the Mendon and Turk-Baker Hill localities. The Junius kames occupy lower ground and show less relief. They are piled on drumlin territory so that kames and drumlins are often confused.

The Waterloo and Auburn moraines show contemporaneous relationships with attenuated drumlin fluting on the north. It appears that the entire extended moraine was contemporaneous with the final shaping of the drumlins in the adjacent territory to the north. Another interesting fact in this connection is the relative weakness of the moraines lying in front of the drumlins with stream outwash largely concentrated in the kame areas. It was suggested that the greater load of drift borne by the ice was mostly incorporated in the drumlins. This belt of kames represents a great volume of drainage which has no apparent genetic relation to the topography. Conditions of the ice, surficial or internal, seem to have determined the concentrated stream flow.

At Mendon Ponds there is an intimate genetic relation between kames and esker, both apparently being a product of the same stream. The kame knolls are outwash detritus deposited where the glacial streams emerged from the fluctuating edge of the ice.

The esker comprises the coarse material dropped by the glacial stream in its bed when the volume and velocity of the flowing water was unable to carry all the load.

The Pinnacle Hills

Location and extent

The Pinnacle Hills (or Pinnacle Range) lie at the eastern portion of the Rochester-Albion moraine. The range consists of an irregular but linear belt of kame deposits extending about 4 miles from the town of Brighton, adjoining Rochester on the southeast, westward to the Genesee River. The range attains a maximum elevation of 749 feet at the Pinnacle Hill high point, and projects above the Rochester plain about 240 feet. The line of hills displays some curvature with a convexity towards the south. The moraine continues beyond the Genesee River northwest toward Spencerport, continuing past Brockport, Holly, and Albion. This western extension of the Rochester-Albion moraine is essentially a belt of subdued hills, knolls, and ridges. In some cases it is present only as ground moraine. East of the Irondequoit Valley is one of the largest drumlin fields in the United States upon which is believed to lie an eastern extension of the moraine. In the drumlin area it is represented by scattered and inconspicuous morainal material.

The earliest description of the Pinnacle Range described the hills as a large kame deposit. The Pinnacle Range was described later as an esker deposited in a "deep ice-walled gorge". The Pinnacle kames after careful study were described as ice-contact deposits laid down by streams discharging from the ice front into glacial Lake Drula. The hypothesis that the Pinnacle Range was an interlobate moraine was later found to be a more accurate depiction.

The Pinnacle Hills have been divided into three main groups which are briefly characterized as follows:

Brighton-Cobbs Hill:

This division encompasses that portion of the range east of Monroe Avenue, including Cobbs Hill. The maximum elevation for this division is 663 feet.

The Pinnacle Hill

The second division is more restricted laterally and extends from Monroe Avenue to about 1/4 mile west of South Clinton Street. It contains the Pinnacle Hill high point at an elevation of 749 feet, approximately 240 feet over the Rochester plain.

The remainder of the range to the Genesee River is included in the third division. It embraces the knoll east of Goodman Street (on which the Colgate-Rochester Divinity School is located), Highland Park between Goodman and South Avenue, Mount Hope Cemetery, and Oak Hill. The latter is the site of The University of Rochester. Summits in the cemetery attain a maximum elevation for this division of 675 feet.

The third division differs from the first two in having a greater width and an outstanding irregular kame-kettle topography similar in form to the Mendon kames and kettles. The dissimilarity between the eastern and western ends may be attributed to different relations between melt water discharge and the configuration of the ice front.

General Description

The northern slopes of the range are steep and irregular probably as a result of both erosion and the direct effects of ice contact. Extensive real estate development and large man made excavations have contributed to the removal or concealment of great quantities of material. Southern slopes are generally less steep and merge into the Lake Dana plain with gentle acclivity (**Figure 17**).

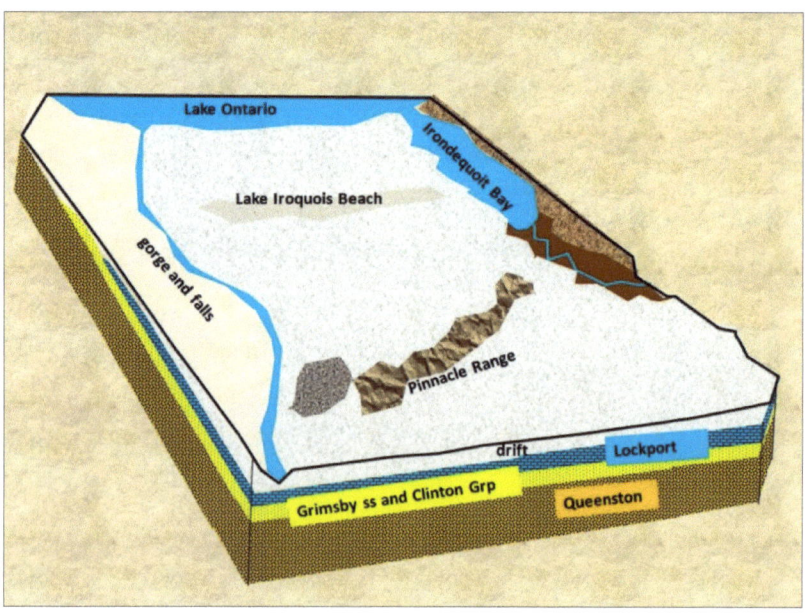

Figure 17. Block diagram showing relationship of major topographic features in immediate vicinity of Rochester.

Water-laid deposits

The Pinnacle Range is composed largely of sand and gravel deposits which display abrupt changes both vertically and laterally. Stratified deposits trend across the range, rather than along its length as would be the case with an esker. Beds dip to the southeast, south, and southwest, the latter taking a slight precedence over the other directions.

Fine sands are found in the eastern and western extremities of the moraine. Fine sands are also found on the north and south slopes with greater abundance of coarser material on the north side. The coarse water-laid materials are of pebble and cobble size. These size ranges are made up of about 50 to 60 percent red Medina sandstone, and 20 to 30 percent Lockport dolomite. The remainder is composed of Silurian-Ordovician limestones and exotic crystallines. Well-rounded pebbles & cobbles of Medina sandstone and Lockport dolomite indicate only a short distance of transport was necessary to produce a well-rounded rock.

Till

An important feature of the Pinnacle Range is the till capping which overlies the stratified sediments and varies in thickness from 3 to 20 feet. The overlying till indicates a moderate southward oscillation of the ice which overrode the kames and coated the hills. Upturned and vertical bedding planes in the stratified material are present on the north side of the hills. High angle faulting in both the till and stratified deposits have been attributed to subsequent slumping after the ice backed away to the north. The south slopes show little distortion in comparison to the northern slopes.

A recent till fabric study shows that the orientation of the long axes of pebbles incorporated in the till varies between southeast and southwest, with the southwest direction showing slightly the greater concentration. Poorly rounded, striated, and polished pebbles and cobbles of Lockport dolomite constitute 60 to 70 percent of the rock fragments in this size range. Lesser and varying amounts of Medina sandstone, Ordovician-Silurian limestones and crystallines make up the other rock types. Large Lockport and crystalline erratics are moderately numerous in the upper till zone.

Till matrix varies between hard compact clays and fine clayey sands or silts. The compact clay matrices are found on the lower northern slopes and, perhaps, represent basal packing beneath the ice mass.

Tills on the upper slopes show a sand-silt matrix with lesser quantities of clay. The pebbles and cobbles in the upper tills generally show a higher degree of rounding than those of the basal till. Apparently the till in the upper reaches of the range was not subjected to the pressures of a thick ice packing but was reworked to some extent by minor melt water action.

Glacial Striae

Glaciated surfaces of the bedrock in the Rochester plain show two sets of striae. The direction of the most prominent set varies between S40W to S60W. Another and less outstanding set of glacial markings display a radiating direction perpendicular to the Pinnacle Ranges West of the Genesee River, the latest ice movement was S5W to S15°W as shown by the striations. Striations recently observed in a new exposure in the city of Rochester have the unexplained and anomalous bearing of S70E.

The two-lobe or interlobate hypothesis

Close to Mount Hope cemetery a subdued till ridge extends in a direct course south to Henrietta. The ridge was indicated as the marker for the ice border which protruded as a lobe south to southwest out of the Irondequoit Valley. The Irondequoit lobe formed a southern ice wall to the Pinnacle Range. Although the till ridge does exist, there is some question as to when it was built relative to the formation of the Pinnacle moraine. Thick clays and silt deposits behind and in front of the ridge are attributed to Lake Dana deposition.

The interlobate interpretation has been put forth on the basis of the form and distribution of the hills, the ice contact deposits on both sides of the range, the considerable height of the range relative to its width, and the drift ridge which extends southward from the west end of the Pinnacle Hills. It is hoped that sedimentological studies now under way may throw additional light on the origin of the hills (**Figure 18**).

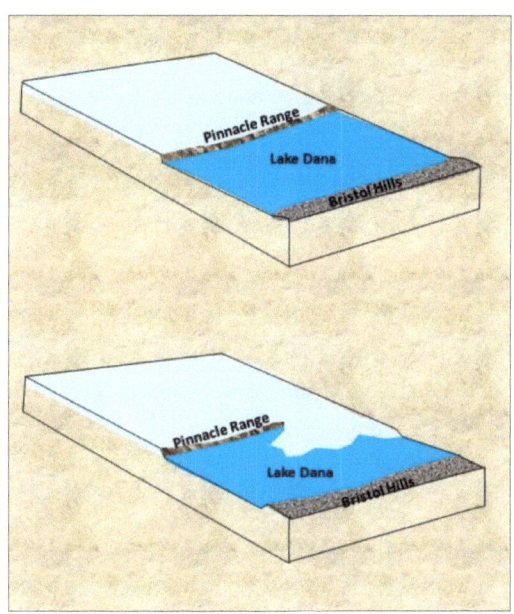

Figure 18. Formation of the Pinnacle Range. Top- single lobe theory; bottom- interlobate theory.

Chapter 5. Letchworth State Park Geology

The New York State Museum provided an overview of the park.

Letchworth Gorge, in Letchworth State Park, is sometimes called "The Grand Canyon of the East". Located in western New York southwest of Rochester, the gorge is best known for the nearly vertical cliffs up to 550 feet and higher which bound the Genesee River.

Three large water falls occur within the canyon, but tens of additional falls occur along the river and along its tributaries in the park. The highest of these, Inspiration Falls, has a total drop of 350 feet, making it the highest waterfall in New York.

Geological Features

The rock seen in the walls of Letchworth Gorge consist of shales, siltstones, and sandstones that were deposited in a shallow sea that covered much of the eastern U.S. during the late Devonian Period. A volcanic ash layer in shales near the Mount Morris Dam, at the lower end of the gorge has been dated at 381 million years ago.

Rocks in the lower part of the gorge consist chiefly of black to gray shale (West River, Middlesex, Chashaqua, and Rhinestreet Formations). Upstream, progressively younger rocks are exposed which feature increasing amounts of siltstone and sandstone belonging to the Angola, Gardeau, and Nunda Formations. The rocks exposed through Letchworth Gorge were deposited as muds to sands over roughly three to five million years of time.

The physical character of Letchworth Gorge varies along its length. The upper and lower sections of the gorge are narrow and deep, cut into bedrock. In contrast, the middle section is a still deep but broad, relatively flat bottomed valley, variously bounded along its walls by either bedrock or glacial sediments (**Figure 19**).

Figure 19. View of Letchworth Gorge from an overlook along the Genesee River. Source: University of Buffalo Department of Geology.

These differences reflect areas where the Genesee River cut new gorges through the rock versus where it flows through an older section of the gorge. The steep and narrow newer sections in the upper and lower reaches have been eroded out since glacial ice retreated from the area prior to approximately 13,000 years ago. The middle section runs through a section of an older gorge in which thick glacial sediments were eroded out after the last glacial retreat. Other parts of the older gorge where the river no longer flows remain fully to partially filled with new glacial till and lake sediments.

Letchworth State Park contains a remarkable exposure of the Frasnian stratigraphic section along the 400 foot high cliffs and waterfalls carved by the Genesee River. Three major waterfalls are viewable in the park, the Lower, Middle and Upper Falls.

Whereas much of the cliff sections are comprised of Gardeau Formation, shales and silty shales interbedded with thin turbidite/storm sandstones, the stratigraphic section above the Upper Falls exposes outcrops of the Nunda Formation. The Nunda Formation is correlative to Elk reservoir sands, and is a good example of a thick turbidite sandstone; however dewatering structures and wave-modification provide more variability than the original Bouma model.

The outcrop exposed along the trail toward the Upper Falls and the High Bridge are described. The path and stairs were cut through 24 m of stratigraphic section, mostly fine-grained sandstone. In addition to examples of primary bed forms and sedimentary structures typical of turbidities, soft-sediment deformation and liquefaction features including sandstone dikes, roll ups, load casts, and zones of liquefaction with sharp boundaries are quite common in both the sandstones and the interbedded shales. Small down-on-the-east stratigraphic offsets (6 cm) along northwest-trending (317 inch) fractures typically coincide with observable deformation zones in the outcrop. Bedding within some of the thicker sandstones appears similar to swaley-cross stratification, which suggests storm-wave modification/influence.

Storm influence within the Nunda Formation is consistent with observations made higher in the stratigraphic section where "deep" water depositional environments (black shales and turbidites) show numerous bed forms (such as hummocky cross-stratification, 3-D ripples, linguoid ripples) that indicate depositional depth was within storm-wave base.

The specific depositional environment of the thick sandstones of the Nunda exposed at Letchworth is ambiguous. The thick nature of the sandstone beds, the striations on the base of the sandstone beds, the massive to planar (to possibly swaley bedding) all indicate proximal turbidites. Such turbidites were typically are thought to represent submarine channel deposits. However, the lack of coarse basal sediments, and the constant thickness of the thick beds exposed along the High Cliffs at Inspiration Point suggest that either the channels were very wide, or that these represent a non-point source, more on the order of a storm-generated wash with a shore face source.

To the east, an erosional channel wall was discovered, indicating a channel (or slump scar), and to the west, a relatively thick Nunda sand bed represented sand lobe at the end of a channel, so it is possible that these sands at Letchworth do represent wide channels (or storm-generated wash that funneled into channels downslope).

At the base of the lower falls, approximately 55 feet of Queenston Formation is present. Thickness of the unit reaches a maximum of 1000 feet. The near complete record of the overlying Silurian Clinton Group can be observed in all three falls in the gorge. In the overlying Lockport Group, fossils are not as abundant as they are elsewhere (**Figures 20 & 21**).

The Salina Group is not represented within the gorge due to the sediments being easily eroded from the landscape. In addition, the Devonian Period sediments are not very well represented in this area due to sedimentary erosion. The Onondega Formation separates the Salina Group from the overlying Hamilton Group. Situated above the Hamilton Group, the Sonyea and West Falls Groups are exposed in the gorge.

At the southern parts of the gorge, progressively younger formations are seen in the walls due to the gentle southward dip of the beds. All three falls within the gorge are capped by resistant sandstone beds within the West Falls Group. Only the lower part of the gorge is seen within the walls.

Figure 20. The Queenston Formation is recognized by its earthy red shale interbedded with reddish to orange yellow sandstone beds. It weathers into shaly fragments. It is Upper Ordovician in age, exposed by about 55 feet. It can be viewed at the Lower Falls in the park (background). Source: Mark Papke, posted on the internet.

The Nunda Sandstone is seen to cap the upper falls and many smaller falls to the south. This region was mainly named for its building stone called bluestone, which some old abandoned quarries may be observed. Between the lower and middle falls, the canyon is occupied by a massively bedded unit consisting of the Lockport Dolomite, a very resistant unit that lies above the Clinton Group (**Figure 22**).

Figure 21. In the lower falls, the Clinton Group overlies the Queenston Formation. In this photo, there are three distinct units present. The lower most unit lies beneath the falls. This may be the Thorold sandstone and siltstone which makes up 5 feet of the lower falls. Overlying the falls and river unit is the Reynales Formation gray calcareous beds which were eroded during high stands of the river level (gray collapsed section at the base of the cliff face). They are overlain by the Rochester Shale, 85 feet of brown to blue gray lower weak shale covered by an upper massive and slightly dolomitic bed. Source: World of Waterfalls posted on the internet.

Figure 22. The Lockport Dolomite unit occupies the steep cliff walls of the section between the lower and middle falls. The rock is a cliff former because dolomite is very resistant to erosion.

However, at the bend in the Genesee River, there is a talus slope (left side) developed in rock which has tumbled down the cliff face and accumulated at the base of slope. It is occupied by trees which established itself in broken fines within the talus pile. On the right side, in the foreground, another talus pile is covered by trees. The dark spots in the lower section of the right side cliff face represent cavities and caves filled in by sediment prior to the overlying layers being deposited. Source: New York State Museum.

Figure 23. Close up view of the Lockport dolomite cliff and talus piles where trees were established where finer grained sizes allowed roots to penetrate the rock pile and water was allowed to seep within the pore spaces (left). At the outcrop level, cross bedding was observed in the limestone, upper right. Lower right shows large fractures developed in a portion of the exposure.

Figure 24. Inspiration Point overlooks the middle falls. Thick bedded Nunda sandstone occurs in the foreground cliff face below the layered beds of the Hatch Formation in upper cliff face. On the right, Nunda sandstone occurs where the green moss starts about mid outcrop. The layered unit above belongs to the Hatch Formation. These units occur above the elevation of the upper falls. Source: Flikr posted on the internet.

Figure 25. The upper falls is capped by the Nunda Sandstone unit. It forms the cliffs on the right and left in the foreground photo. The layered unit below the Nunda sandstone belongs to the Gardeau Formation. Thickness is about 200 feet in this location. Source: Stav is lost posted on the internet.

Figure 26. River cut bank into Gardeau Sandstone just below the upper falls. Source: Jay Greenburg posted on the internet.

Figure 27. Situated between the Gardeau and Hatch Formations, the Nunda Sandstone unit separates the two formations. This photo is a close up view of the Nunda sandstone unit. Source: University of Buffalo posted on the internet.

Figure 28. Tributary creek flowing over Nunda sandstone near the upper falls. Source: Loaded landscapes posted on the internet.

References

Jones, M., and Jones, M. ___. Letchworth State Park. St. Lawrence University Geology Club. Published on the internet and references contained within.

Smith, G.J., and Jacobi, R.D., 2006. Depositional and Tectonic Models for Upper Devonian Sandstones in Western New York State in Guidebook for the field trip held October 7th, 2006 in conjunction with the 35th Eastern Section AAPG Meeting and 78th NYSGA Field trips held in Buffalo, New York. UB Rock Fracture Group, Department of Geology, University at Buffalo, Buffalo, New York

The New York State Museum. 2019. Letchworth Go: An overview of the Grand Canyon of the East.

The University of Rochester, 1956. Guidebook. Twenty-eighth Annual Meeting of the New York State Geological Association. Department of Geology and Geography at the University of Rochester, NY and references contained within.

www.ingramcontent.com/pod-product-compliance
Lightning Source LLC
Chambersburg PA
CBHW040226220526
45473CB00001B/137